Communications in Computer and Information Science 2807

Rationale

The CCIS series is devoted to the publication of proceedings of computer science conferences. Its aim is to efficiently disseminate original research results in informatics in printed and electronic form. While the focus is on publication of peer-reviewed full papers presenting mature work, inclusion of reviewed short papers reporting on work in progress is welcome, too. Besides globally relevant meetings with internationally representative program committees guaranteeing a strict peer-reviewing and paper selection process, conferences run by societies or of high regional or national relevance are also considered for publication.

Topics

The topical scope of CCIS spans the entire spectrum of informatics ranging from foundational topics in the theory of computing to information and communications science and technology and a broad variety of interdisciplinary application fields.

Information for Volume Editors and Authors

Publication in CCIS is free of charge. No royalties are paid, however, we offer registered conference participants temporary free access to the online version of the conference proceedings on SpringerLink (http://link.springer.com) by means of an http referrer from the conference website and/or a number of complimentary printed copies, as specified in the official acceptance email of the event.

CCIS proceedings can be published in time for distribution at conferences or as post-proceedings, and delivered in the form of printed books and/or electronically as USBs and/or e-content licenses for accessing proceedings at SpringerLink. Furthermore, CCIS proceedings are included in the CCIS electronic book series hosted in the SpringerLink digital library at http://link.springer.com/bookseries/7899. Conferences publishing in CCIS are allowed to use our online conference service (Meteor) for managing the whole proceedings lifecycle (from submission and reviewing to preparing for publication) free of charge.

Publication process

The language of publication is exclusively English. Authors publishing in CCIS have to sign the Springer CCIS copyright transfer form, however, they are free to use their material published in CCIS for substantially changed, more elaborate subsequent publications elsewhere. For the preparation of the camera-ready papers/files, authors have to strictly adhere to the Springer CCIS Authors' Instructions and are strongly encouraged to use the CCIS LaTeX style files or templates.

Abstracting/Indexing

CCIS is abstracted/indexed in DBLP, Google Scholar, EI-Compendex, Mathematical Reviews, SCImago, Scopus. CCIS volumes are also submitted for the inclusion in ISI Proceedings.

How to start

To start the evaluation of your proposal for inclusion in the CCIS series, please send an e-mail to ccis@springer.com

Emilio Corchado · Héctor Quintián ·
Hilde Pérez García · José Luis Calvo Rolle ·
Sérgio Filipe Ramos ·
Francisco Javier Martínez de Pisón ·
Álvaro Herrero Cosío · Paolo Fosci
Editors

Computational Intelligence in Security for Information Systems

18th International Conference, CISIS 2025
Salamanca, Spain, October 16–17, 2025
Proceedings

Editors
Emilio Corchado
University of Salamanca
Salamanca, Spain

Hilde Pérez García
University of León
León, Spain

Sérgio Filipe Ramos
Polytechnic Institute of Porto
Porto, Portugal

Álvaro Herrero Cosío
University of Burgos
Burgos, Spain

Héctor Quintián
University of A Coruña
Ferrol, La Coruña, Spain

José Luis Calvo Rolle
University of A Coruña
Ferrol, La Coruña, Spain

Francisco Javier Martínez de Pisón
University of La Rioja
Logroño, Spain

Paolo Fosci
University of Bergamo
Bergamo, Italy

ISSN 1865-0929 ISSN 1865-0937 (electronic)
Communications in Computer and Information Science
ISBN 978-3-032-19769-6 ISBN 978-3-032-19770-2 (eBook)
https://doi.org/10.1007/978-3-032-19770-2

This Springer imprint is published by the registered company Springer Nature Switzerland AG
The registered company address is: Gewerbestrasse 11, 6330 Cham, Switzerland

Preface

This volume of Communications in Computer and Information Science contains accepted papers presented at the *18th International Conference on Computational Intelligence in Security for Information Systems* (CISIS 2025), which was held in the beautiful city of Salamanca, Spain, in October 16–17, 2025.

The CISIS 2025 conference aimed to provide a meeting opportunity for academic and industry-related researchers from various communities in Computational Intelligence, Information Security, and Data Mining. The need for intelligent, flexible behaviour by large, complex systems, especially in mission-critical domains, was intended to be the catalyst and the aggregation stimulus for the overall event.

After a single-blind peer-review process in which the 36 submissions each received an average of three reviews, the CISIS 2025 International Program Committee selected 17 papers for publication in these conference proceedings, resulting in an acceptance rate of 47%. In this edition, one special session was organised: Artificial Intelligence for Protecting the Internet of Things.

The selection of papers was extremely rigorous to maintain the high quality of the conference. We want to thank the members of the Program Committees for their diligent work during the review process. This is a crucial process for creating a high-standard conference; the CISIS series of conferences would not exist without their help.

CISIS 2025 enjoyed outstanding keynote speeches by distinguished guest speakers: Ajith Abraham at Sai University (India) and Sung-Bae Cho at Yonsei University (South Korea).

CISIS 2025 has teamed up with "*Logic Journal of the IGPL*" (Oxford University Press) for a suite of special issues, including selected papers from CISIS 2025.

Particular thanks go as well to the conference's main sponsors, Startup Olé, the CIBER-OLÉ project (within the National Cybersecurity Industry Promotion Program, framed within the INCIBE Emprende program and financed by INCIBE and the University of Salamanca), the BISITE research group at the University of Salamanca, the CTC research group at the University of A Coruña, and the University of Salamanca. They jointly contributed actively and constructively to the success of this initiative. This activity was carried out in execution of the Strategic Project "Critical infrastructures cybersecure through intelligent modeling of attacks, vulnerabilities and increased security of their IoT devices for the water supply sector" (C061_/23), the result of a collaboration agreement signed between the National Institute of Cybersecurity (INCIBE) and the University of A Coruña. This initiative was carried out within the framework of the funds of the Recovery, Transformation and Resilience Plan, financed by the European Union (Next Generation), the project of the Government of Spain that outlines the roadmap for the modernization of the Spanish economy, the recovery of economic growth and job creation, for solid, inclusive and resilient economic reconstruction after the COVID19 crisis, and to respond to the challenges of the next decade. This activity was also promoted by PID2022-137152NB-I00 funded by MICIU/AEI/10.13039/501100011033 and by ERDF/EU.

We would like to thank all the special session organizers, the contributing authors, the Program Committees, and the Local Organizing Committee members for their hard and highly valuable work, which contributed to the success of the CISIS event.

The editors

October 2025

Emilio Corchado
Héctor Quintián
Hilde Pérez García
José Luis Calvo Rolle
Sérgio Filipe Ramos
Francisco Javier Martínez de Pisón
Álvaro Herrero Cosío
Paolo Fosci

Organization

General Chair

Emilio Corchado	University of Salamanca, Spain

Program Committee Chairs

Emilio Corchado	University of Salamanca, Spain
Héctor Quintián	University of A Coruña, Spain
Alicia Troncoso Lora	Pablo de Olavide University, Spain
Hilde Pérez García	University of León, Spain
Esteban Jove	University of A Coruña, Spain
José Luis Calvo Rolle	University of A Coruña, Spain
Francisco Javier Martínez de Pisón	University of La Rioja, Spain
Pablo García Bringas	University of Deusto, Spain
Francisco Martínez Álvarez	Pablo de Olavide University, Spain
Álvaro Herrero Cosío	University of Burgos, Spain
Paolo Fosci	University of Bergamo, Italy
Sérgio Filipe Ramos	Polytechnic Institute of Porto, Portugal

Program Committee

Adam Wójtowicz	Poznań University of Economics and Business, Poland
Agustín García Fisher	University of A Coruña, Spain
Agustín Martín Muñoz	Spanish National Research Council (CSIC), Spain
Alberto Peinado	University of Málaga, Spain
Alejandro Vidal Bralo	University of A Coruña, Spain
Álvaro Michelena	University of A Coruña, Spain
Andrés Pérez Zabala	Basque Research and Technology Alliance, Spain
Andrysiak Tomasz	University of Technology and Life Sciences (UTP), Poland
Ángel Arroyo	University of Burgos, Spain
Antonio Díaz-Longueira	University of A Coruña, Spain
Carlos Pereira	ISEC, Portugal
Ciprian Pungila	West University of Timişoara, Romania

Cristina Alcaraz	University of Málaga, Spain
Daniel Urda	University of Burgos, Spain
Darius Galis	West University of Timisoara, Romania
Eduardo Solteiro Pires	University of Trás-os-Montes and Alto Douro, Portugal
Enrique Onieva	University of Deusto, Spain
Esteban Jove	University of A Coruña, Spain
Fernando Ribeiro	EST, Portugal
Fernando Tricas	University of Zaragoza, Spain
Francisco Martínez-Álvarez	Pablo de Olavide University, Spain
Francisco Zayas-Gato	University of A Coruña, Spain
Hugo Sanjurjo-González	University of Deusto, Spain
Hugo Scolnik	ARSAT SA, Argentina
Ioannis Sorokos	Fraunhofer IESE, Germany
Isaias Garcia	University of León, Spain
Jesús Díaz-Verdejo	University of Granada, Spain
Jose Barata	Universidade NOVA de Lisboa, Portugal
José Aveleira Mata	Research Institute of Applied Sciences in Cybersecurity, Spain
José F. Torres	Pablo de Olavide University, Spain
Jose A. Onieva	University of Málaga, Spain
Jose Carlos Metrolho	Polytechnic Institute of Castelo Branco, Portugal
Jose Luis Calvo-Rolle	University of A Coruña, Spain
Jose M. Molina	University Carlos III of Madrid, Spain
Jose Manuel Lopez-Guede	University of the Basque Country, Spain
José-Luis Casteleiro-Roca	University of A Coruña, Spain
Josep Ferrer	Universitat de les Illes Balears, Spain
Juan J. Gude	University of Deusto, Spain
Juan Jesús Barbarán	University of Granada, Spain
Manuel Graña	University of the Basque Country, Spain
Manuel Rubiños Trelles	University of A Coruña, Spain
Marta María Álvarez Crespo	University of A Coruña, Spain
Michal Choras	Bydgoszcz University of Science and Technology, Poland
Michal Wozniak	Wrocław University of Technology, Poland
Míriam Timiraos Díaz	University of A Coruña, Spain
Pablo García Bringas	University of Deusto, Spain
Paula Patricia Arcano Bea	University of A Coruña, Spain
Petrica Pop	Technical University of Cluj-Napoca, North University Centre of Baia Mare, Romania
Rafael Álvarez	University of Alicante, Spain
Rafael Corchuelo	University of Seville, Spain

Raúl Durán	University of Alcalá, Spain
Roberto Casado-Vara	University of Burgos, Spain
Rudolf Erdei	Holisun SRL, Romania
Salvador Alcaraz	Miguel Hernández University of Elche, Spain
Sorin Stratulat	Université de Lorraine, Metz, France

CISIS 2025: Special Sessions

Artificial Intelligence for Protecting the Internet of Things

Program Committee

Álvaro Herrero (Organizer)	University of Burgos, Spain
Daniel Urda (Organizer)	University of Burgos, Spain
Diego Granados (Organizer)	University of Burgos, Spain
Adam Wójtowicz	Poznań University of Economics and Business, Poland
Darius Galis	West University of Timisoara, Romania
Dominik Olszewski	Warsaw University of Technology, Poland
Michał Choraś	Bydgoszcz University of Science and Technology, Poland
Nuno Alberto Ferreira Lopes	Polytechnic Institute of Cávado and Ave, Portugal
Oliviu Matei	Technical University of Cluj-Napoca, North University Centre of Baia Mare, Romania
Petrica Pop	Technical University of Cluj-Napoca, North University Centre of Baia Mare, Romania

CISIS 2025 Organizing Committee Chairs

Emilio Corchado	University of Salamanca, Spain
Héctor Quintián	University of A Coruña, Spain

CISIS 2025 Organizing Committee

José Luis Calvo Rolle	University of A Coruña, Spain
Esteban Jove	University of A Coruña, Spain
José Luis Casteleiro Roca	University of A Coruña, Spain
Francisco Zayas Gato	University of A Coruña, Spain

Álvaro Michelena	University of A Coruña, Spain
Míriam Timiraos Díaz	University of A Coruña, Spain
Antonio Javier Díaz Longueira	University of A Coruña, Spain
Paula Patricia Arcano Bea	University of A Coruña, Spain
Manuel Rubiños Trelles	University of A Coruña, Spain
Marta María Álvarez Crespo	University of A Coruña, Spain
Alejandro Vidal Bralo	University of A Coruña, Spain
Agustín García Fisher	University of A Coruña, Spain
Iker Pastor López	University of Deusto, Spain

Contents

AI Safety, Privacy and Trustworthy Systems

Special Session: Artificial Intelligence for Protecting the Internet of Things

Malware and Intrusion Detection

A Fast Metric for Preliminary Risk Assessment of Portable Executables Using Static Analysis

Emanuel Ioan Albu(✉) and Ciprian Pungilă

Faculty of Informatics, Department of Digital Technologies and Software Engineering, West University of Timisoara, Timisoara, Romania
{emanuel.albu01,ciprian.pungila}@e-uvt.ro

Abstract. This paper introduces a lightweight, fast, and interpretable static analysis method for the preliminary risk assessment of Windows Portable Executable (PE) files. The proposed system combines three complementary techniques: entropy analysis to detect obfuscation, structural metadata inspection to identify anomalous PE characteristics, and YARA (Yet Another Ridiculous Acronym) rule matching for signature-based threat identification. A custom point-based scoring model, empirically derived from 1,000 benign PE files, assigns risk scores based on statistical deviations in header fields and matched threat patterns. The tool was validated on a balanced dataset of 1,000 PE samples (500 benign, 500 malicious), achieving 80% detection accuracy, an 8% improvement over the ClamAV static scanner, while maintaining perfect precision and significantly reduced analysis time. Its design emphasizes modularity, enabling seamless integration with other tools and supporting the broader trend toward explainable AI in cybersecurity.

Keywords: Static analysis · Malware detection · Portable executable · Malicious · Benign

1 Introduction

Static analysis has become a pivotal approach in malware detection, offering the ability to extract and interpret features from binaries without executing them. This mitigates the risks associated with dynamic analysis while enabling the identification of malicious patterns, control flow anomalies, and embedded indicators of compromise (IoCs). One of the most common formats targeted in static analysis is the Windows Portable Executable (PE) file - a structure ubiquitous in Windows-based systems, encompassing executables, object code, and DLLs. Derived from the Common Object File Format (COFF), the PE structure contains metadata essential for program execution, rendering it a rich source for feature extraction and threat assessment.

This work proposes a fast, modular, and interpretable static analysis framework for the preliminary risk assessment of PE files. The methodology integrates

E. Corchado et al. (Eds.): CISIS 2025, CCIS 2807, pp. 3–12, 2026.
https://doi.org/10.1007/978-3-032-19770-2_1

three complementary components: entropy analysis for detecting obfuscation, structural metadata inspection for identifying anomalous characteristics, and YARA-based rule matching for signature-driven detection. A custom scoring model, empirically derived from a corpus of 1,000 benign PE files, enables a granular quantification of risk through deviations in key fields and rule matches.

In summary, this work proposes a scalable and efficient framework for static malware risk assessment using PE files. Its easy-to-handle structure and focus on prompt identification make it ideal for incorporating into email filters, systems that prevent intrusions, and security tools for endpoints where quick file evaluation is crucial. By prioritizing clear and well-explained functionalities, the system also supports the increasing requirement for understandable artificial intelligence in the field of cybersecurity.

The reaminder of this paper is organized as follows. Section 2 provides an overview of the related work and current techniques in static PE malware detection. Section 3 describes the design and implementation of the proposed methodology, including entropy analysis, structural metadata inspection, and YARA rule integration. Section 4 presents the experimental setup, dataset details, and comparative evaluation with existing tools. Finally, Sect. 5 concludes the paper, summarizing the key findings and outlining potential directions for future research.

2 State of the Art

The detection of malicious Portable Executable (PE) files through static analysis has garnered increasing attention due to its suitability for resource-constrained and high-risk environments. Unlike dynamic analysis, static approaches allow early detection without executing potentially harmful code. This section discusses key advancements in static PE malware detection, categorized by analysis techniques [1].

Several approaches utilize machine learning models that rely on PE header features to distinguish between benign and malicious files. Raff et al. [2] demonstrated that neural networks trained on minimal header information can achieve high detection rates. Similarly, Kumar and Shetty [3] showed that metadata fields like `SizeOfHeaders` and `AddressOfEntryPoint` are predictive indicators when used in classifiers, confirming that header anomalies are a lightweight yet effective malware signal.

Entropy analysis is a well-established technique to detect obfuscation and packing in binaries. Choi et al. [4] proposed the PHAD approach, which utilizes entropy irregularities to flag packed executables. Matin [1] further supported this by identifying elevated entropy as a signature of ransomware and polymorphic malware. While useful, entropy-based methods must account for benign compression to minimize false positives.

Other researchers have focused on structural metadata and anomalies in the PE file layout. Santos and Festijo [5] and Tyagi et al. [8] observed that irregularities in fields such as `SectionAlignment`, `ImageBase`, and `TimeDateStamp` are often indicative of hand-crafted or obfuscated malware. These features are valuable due to their interpretability and resistance to evasion.

YARA has emerged as a dominant tool for identifying known malware signatures within static binaries. Mahdi and Trabelsi [10] demonstrated that YARA rules targeting suspicious APIs, hardcoded IPs, and embedded URLs (Uniform Resource Locators) are effective for detecting Command-and-Control (C2) infrastructure. Kamble and Sridevi [9] highlighted how such signatures, when combined with entropy and header analysis, enhance detection capability while maintaining low false positive rates.

Recent work has focused on hybrid approaches that integrate multiple detection strategies. Ucci et al. [11] surveyed a wide spectrum of machine learning and hybrid models, noting the lack of interpretable systems suitable for real-time deployment. Our approach addresses this gap by proposing a modular, scoring-based detection method that empirically calibrates thresholds using benign data, with the added benefit of explainability and speed.

2.1 Benchmarking and Comparative Tools

ClamAV [6] is a commonly used open-source antivirus engine that relies on signature-based static detection. While effective against known threats, ClamAV has limited capabilities when handling obfuscated or novel malware, making it a useful baseline for evaluating more sophisticated systems. Our experiments confirm that a hybrid, feature-rich scoring model can outperform ClamAV in both speed and accuracy.

3 Methodology Design

The proposed detection script is a custom-built Python tool available publicly in GitHub [7] designed for the static analysis of PE files, as outlined in Fig. 1. The process begins with scanning each file from a folder regardless of whether the file has or not an extension, after which the file is sent to further binary analysis, applicable only to PE files.

The first step involves entropy computation of each section of the executable, in order to assess its obfuscation level. This also helps determine, to a high degree of accuraccy, if the file is packed. The structural features of the PE file are then examined, with these checks also including, for example: the examination of the header size and ensuring it matches a certain range, alignment of sections, the image base address, the entry point address (i.e. is it part of the existing sections), and the time-date stamp.

Next, the workflow applies several precompiled sets of YARA rules. The first set is used for packed executables, looking for known packer signatures, as well as common identifiers. Moving forward, the next set of rules aim to identify suspicious strings commonly found in malware, with further rule checks being conducted in order to look for suspicious dynamic link libraries (DLLs) imports, hard-coded Internet protocol (IP) addresses, as well as embedded URLs. Every match found increases the risk score, activating a specific flag.

The next step involves looks to identify relationships between relevant static indicators of malware. For example, if a binary is packed and contains suspicious strings, or, similarly, if it demonstrates high entropy of the file itself, and for its various sections, it is considered that a match had been found. The same logic applies to packed binaries that contain suspicious/flagged strings and import functions from DLLs that also contain suspicious/flagged strings, determining the flagging of these DLLs as well. The risk score is adjusted accordingly, depending on each of the scenarios matched, as outlined before.

The final risk score is computed as a final step, and a binary classification happens (malicious or non-malicious) based on its value exceeding an empirically-determined threshold. A final report of the resulting analysis process is also displayed.

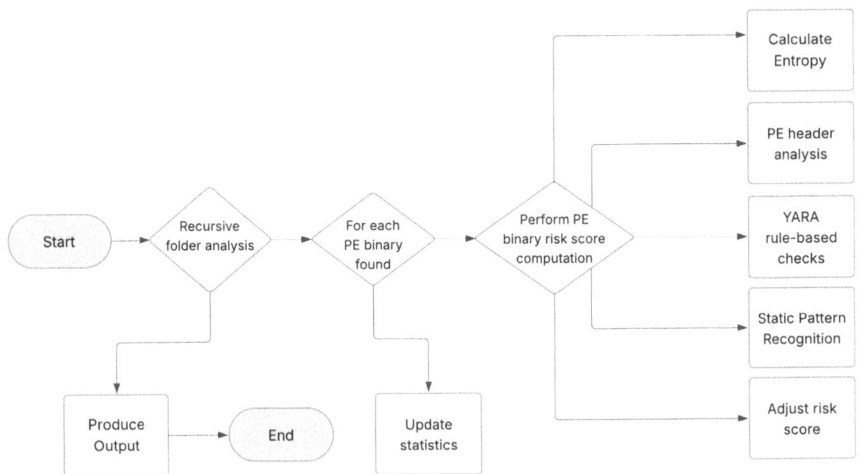

Fig. 1. Script control flow graph.

3.1 Feature Extraction

Feature extraction constitutes a fundamental component of static analysis. The process involves identifying and selecting features based on their relevance in indicating whether a file exhibits behaviors typically associated with malicious software. In our approach, features are extracted from three primary domains: entropy analysis, structural metadata from the PE format, and rule-based detection using YARA signatures.

Entropy functions as a statistical measure of data randomness and is widely utilized in malware analysis to detect obfuscation techniques such as packing, encryption, or compression [4]. Within our system, entropy is calculated at two levels of granularity: the file level and the section level. For both levels, the entire binary content is processed by a function that computes Shannon entropy, based on the distribution of byte values, using the following formula:

$$H = -\sum_{i=1}^{n} p_i \log_2(p_i) \tag{1}$$

3.2 Entropy and Structural PE Metadata Analysis

The variable p_i denotes the probability of occurrence of the i-th byte value, while n represents the number of unique byte values observed within the data. Files exhibiting low entropy tend to be uncompressed and well-structured. Conversely, files approaching the theoretical maximum entropy value of 8.0 (for byte-level data) often contain encrypted payloads or compressed content, which are typical traits of malicious binaries.

Structural characteristics of PE files provide valuable indicators for distinguishing benign executables from obfuscated malware [11]. Our system uses the `pefile` library to extract features from the PE and optional headers, including `SizeOfHeaders`, where atypical values outside 0x200–0x1000 may suggest tampering [3]. The `SectionAlignment` field, normally $\geq$ 0x1000 in legitimate binaries, raises suspicion when below 0x200, indicating possible manual crafting or obfuscation [1]. The `ImageBase`, typically between 0x400000 and 0x80000000, can signal custom compilers or memory-staging shellcode when it deviates [8]. `AddressOfEntryPoint` should reside in standard sections like `.text`; if outside mapped regions, it may imply injected shellcode or a malformed header [1]. Finally, `TimeDateStamp` often reflects realistic build times in benign software, whereas malware may falsify this field with zeros or implausibly old values to evade detection.

3.3 Signature-Based Detection with YARA

To augment static analysis with pattern-based detection, our system integrates **YARA**, a rule-based tool designed for identifying malware through string and binary pattern matching. We apply five YARA rule sets to each file: a) known packers, b) suspicious string patterns, c) DLL imports, d) hard-coded IP addresses, and e) embedded URLs. These rules are precompiled for performance and applied using the `yara-python` library. Each rule consists of string patterns and logical conditions designed to reflect common malware traits. The **packer rules** detect well-known packing utilities such as UPX, ASPack, and Themida. The **string rules** search for indicative content such as obfuscated commands and exploit-specific keywords. The **DLL import rules** emphasize the use of critical APIs like `VirtualAlloc`, `WriteProcessMemory`, `CreateRemoteThread`, and `LoadLibrary`, which are often leveraged in memory manipulation and process injection [9]. APIs from libraries such as `kernel32.dll`, `ntdll.dll`, `ws2_32.dll`, and `wininet.dll` are considered high risk due to their association with process creation, memory allocation, and network communications. The **IP and URL rules** identify statically embedded addresses that could facilitate Command-and-Control (C2) communication or data exfiltration [10]. Such hardcoded network indicators are a common trait of advanced persistent threats (APT). All YARA rules employed in this research are sourced from a publicly available GitHub repository maintained by security professionals and the threat intelligence community [12]. These rules provide a rich source of expert-curated signatures, increasing our system's capability to detect obfuscated or compressed malware samples.

3.4 Scoring System

To assess the risk associated with a binary, we employ a dynamic and robust scoring system grounded in empirical testing. Specifically, we analyzed 1,000 benign executable files sourced from standard Windows system directories. These samples served to benchmark the normal distribution of relevant features and to define precise scoring criteria. We selected these Windows files due to their practical relevance: although benign, they may share certain structural or functional characteristics with malware-such as the inclusion of libraries like `kernel32.dll`, `CertEnroll.dll`, and `ntdll.dll`.

Threshold Definition. To determine a scoring threshold, we computed the average risk score across the 1,000 benign files and recorded the highest individual score. We then defined the "sweet spot" threshold as:

$$\text{Threshold} = \text{Average Risk Score} + 0.5 \times \text{Max Risk Score}$$

Feature-Based Scoring. As illustrated in Table 1, our core logic evaluates how frequently each feature occurs in the benign dataset. The score for each feature begins at 10 points and is calculated using the formula:

$$\text{score} = 10 \times x \tag{2}$$

Here, x is the proportion of Windows files that **do not** exhibit the respective feature. This approach is applied to features derived from structural PE metadata, including `SizeOfHeaders`, `SectionAlignment`, `ImageBase`, as well as `AddressOfEntryPoint` and invalid or unusual `TimeDateStamp`.

Entropy-Based Scoring. Following our analysis of the 1,000 benign files, we observed an average file-level entropy of approximately 6. To reflect increasing uncertainty, we implemented an exponentially growing scoring function for entropy values exceeding 5.5. The scoring function $S(H)$ is defined as:

Table 1. Distribution of atypical features in 1,000 benign Windows executables.

Feature	Have	Don't have
SizeOfHeaders	0	1.000
SectionAlignment	0	1.000
ImageBase	140	860
AddressOfEntryPoint	300	700
TimeDateStamp	20	980

$$S(H) = \begin{cases} 0, & \text{if } H < 5.5 \\ S_{\max} \cdot 0.4 \cdot \dfrac{b^{H/p} - 1}{b - 1}, & \text{if } 5.5 \leq H \leq p \\ S_{\max} \cdot 0.4 + S_{\max} \cdot 0.6 \cdot \dfrac{b^{\frac{H-p}{H_{\max}-p}} - 1}{b - 1}, & \text{if } p < H \leq H_{\max} \end{cases}$$

where:

- H is the observed entropy,
- p is the pivot point (set at the average entropy of 6),
- $H_{\max}$ is the maximum theoretical entropy (8.0),
- b is the base of exponential growth,
- $S_{\max} = 20$ is the maximum score assigned for entropy.

Entropy values below 5.5 receive a score of 0 (Fig. 2).

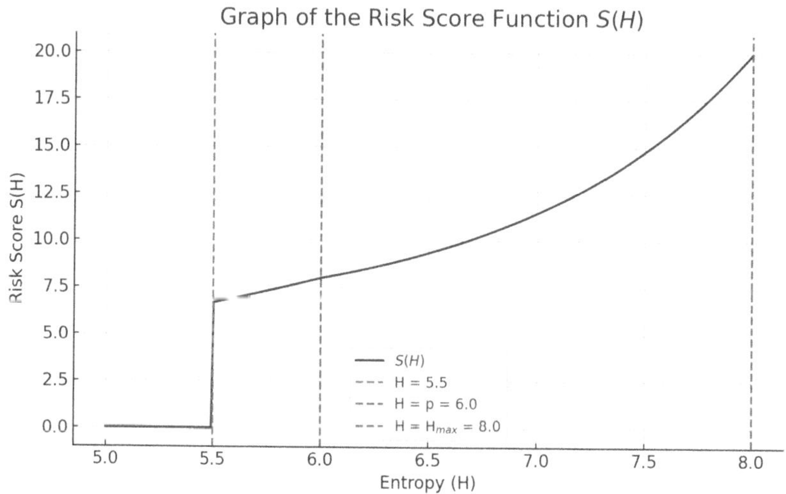

Fig. 2. Graph of the risk score function.

YARA Rule Scoring. YARA rules are treated as discrete, high-impact indicators. Each rule match adds 10 points to the final score. The system evaluates matches against five YARA rule categories: a) known packers, b) suspicious strings, c) DLL imports, e) hard-coded IP addresses, and f) embedded URLs. The final YARA score is computed as:

$$\text{YARA Score} = 10 \times \text{Number of Matching Rules}$$

Pattern Matching Heuristics. To further refine detection accuracy, we incorporate a heuristic pattern-matching mechanism. This system identifies combinations of features that frequently co-occur in malware samples and assigns an additional score of 20 points to such patterns. Specific combinations that trigger this boost include: a) packed binary and any of {suspicious strings, high entropy (file or section level), suspicious strings + DLL imports}, and b) suspicious strings + DLL imports.

4 Experimental Results

The dataset consists of a total of 1,000 PE files, equally divided into 500 benign and 500 malicious samples. Malicious samples were obtained through collaboration with the administrators of VirusShare [13], a well-established malware repository widely used in the cybersecurity research community. These malicious binaries represent a diverse collection, including packed malware, trojans, downloaders, and other types of threats. To construct the benign portion of the dataset, 500 PE files were collected from a clean Windows environment. These files include system utilities, administrative tools, and default applications typically found in genuine installations of the Windows operating system. A batch script was used to automate the extraction of these binaries. Using native Windows files as benign samples offers a meaningful baseline, as they often share certain structural similarities with malicious binaries.

A VirtualBox [14] virtual machine (VM) was configured for controlled testing, featuring 8 GB of memory, 6 logical processors, and a 100 GB virtual disk running Windows 11. To ensure a secure and isolated environment, all non-essential services were disabled, no third-party applications were installed, and features such as shared folders, clipboard sharing, and drag-and-drop were deactivated. After transferring the malicious files, the network was disconnected and Windows Defender was disabled. The 1,000 binaries were organized into two separate folders—one for benign files and one for malicious samples. The analysis system was executed independently on each folder, and the results were automatically saved to structured text files for evaluation. To validate the accuracy and reliability of our system, we integrated ClamAV [6], an open-source antivirus engine focused exclusively on static analysis. ClamAV was executed using its command-line interface via a custom batch script, which also recorded execution times and detection results.

The comparative evaluation between our proposed static analysis system and ClamAV reveals notable differences in performance, as illustrated in Fig. 3. Our system achieved an 80% detection rate for malicious files, outperforming ClamAV's 72% detection rate. Both systems achieved a precision of 100%. Execution time analysis showed that our system performed significantly faster. This improvement is attributed to the targeted feature extraction methodology and the use of precompiled YARA rules, which streamline the static analysis process.

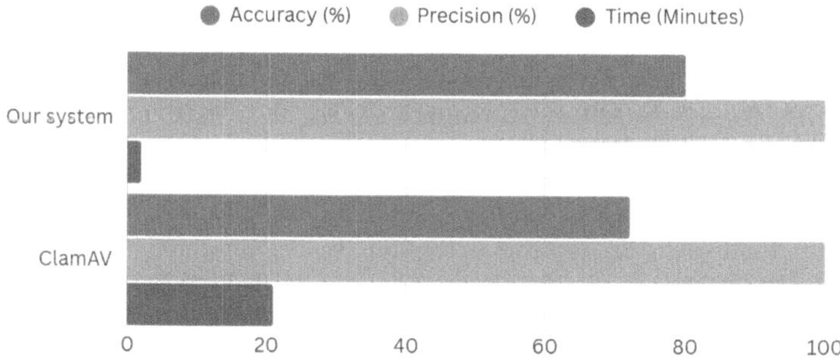

Fig. 3. Experimental results of the proposed static analysis approach, as compared to ClamAV.

5 Conclusions

Experimental validation demonstrates that the proposed method not only achieves high detection accuracy but also significantly outperforms traditional tools such as ClamAV in terms of speed and precision. With its lightweight design and explainable architecture, the system is well-suited for deployment in time-sensitive environments such as email filters, endpoint security tools, and intrusion prevention systems.

Future work will aim to refine the scoring system by incorporating a broader set of heuristics and expanding YARA rules to cover emerging malware patterns and advanced obfuscation techniques. Additional static features, such as PE section anomalies, import/export table irregularities, and certificate metadata, will be explored to improve detection granularity. Efforts will also focus on optimizing rule prioritization and weighting within the scoring model to further enhance accuracy while maintaining the system's speed and interpretability.

Acknowledgments. This work has been partially supported by (1) the project RoNaQCI, part of EuroQCI, DIGITAL-2021-QCI-01- DEPLOY-NATIONAL, 101091562, (2) the project "Romanian Hub for Artificial Intelligence - HRIA", Smart Growth, Digitization and Financial Instruments Program, 2021–2027, MySMIS no. 351416.

References

1. Matin, I.M.M.: Ransomware extraction using static portable executable (PE) feature-based approach. In: 2023 6th International Conference of Computer and Informatics Engineering (IC2IE), Lombok, Indonesia, pp. 70–74 (2023). https://doi.org/10.1109/IC2IE60547.2023.10331246. Accessed 16 Apr 2025
2. Raff, E., Sylvester, J., Nicholas, C.: Learning the PE header malware detection with minimal domain knowledge. In: Proceedings of the ACM Workshop on Artificial Intelligence and Security, pp. 121–132 (2017). Accessed 27 Apr 2025
3. Kumar, S.S., Shetty, J.: Malicious PE file detection using machine learning: an analysis of header features. In: Proceedings of COSMIC, pp. 66–71 (2024). Accessed 27 Apr 2025
4. Choi, Y.-S., Kim, I.-K., Oh, J.-T., Ryou, J.-C.: PE file header analysis-based packed PE file detection technique (PHAD). In: International Symposium on Computer Science and its Applications, pp. 28–31 (2008). Accessed 22 Apr 2025
5. Santos, R.S., Festijo, E.D.: Generating features of windows portable executable files for static analysis using portable executable reader module (PEFile). In: IC2IE, pp. 283–288 (2021). Accessed 27 Apr 2025
6. ClamAV. https://www.clamav.net. Accessed 22 Apr 2025
7. Script source code. https://github.com/emichulo/Automation-Script-using-Static-Analysis-For-PE
8. Tyagi, S., Baghela, A., Dar, K.M., Patel, A., Kothari, S., Bhosale, S.: Malware detection in PE files using machine learning. In: 2022 OPJU International Technology Conference on Emerging Technologies for Sustainable Development (OTCON), Raigarh, Chhattisgarh, India, pp. 1–6 (2023). https://doi.org/10.1109/OTCON56053.2023.10113998. Accessed 20 Apr 2025
9. Kamble, M.T., Sridevi: Feature extraction and analysis of portable executable malicious file. In: 2022 Second International Conference on Computer Science, Engineering and Applications (ICCSEA), Gunupur, India, pp. 1–6 (2022). https://doi.org/10.1109/ICCSEA54677.2022.9936121. Accessed 20 Apr 2025
10. Mahdi, R.H., Trabelsi, H.: Detection of malware by using YARA rules. In: 2024 21st International Multi-Conference on Systems, Signals & Devices (SSD), Erbil, Iraq, pp. 1–8 (2024). https://doi.org/10.1109/SSD61670.2024.10549308. Accessed 20 Apr 2025
11. Ucci, D., Aniello, L., Baldoni, R.: Survey of machine learning techniques for malware analysis. Comput. Secur. **81**, 123–147 (2019). ISSN 0167–4048. https://doi.org/10.1016/j.cose.2018.11.001. Accessed 27 Apr 2025
12. Repository of yara rules. https://github.com/Yara-Rules/rules. Accessed 22 Apr 2025
13. VirusShare.com. https://virusshare.com. Accessed 22 Apr 2025
14. Oracle VirtualBox. https://www.virtualbox.org. Accessed 20 Apr 2025

Unsupervised Online Learning for Network Flow Anomaly Detection: A Comparative Evaluation

Alberto Miguel-Diez(✉), Claudia Álvarez-Aparicio, Adrián Campazas-Vega, Vicente Matellán-Olivera, and Ángel Manuel Guerrero-Higueras

Robotics Group, University of León, MIC building, Campus de Vegazana, s/n, 24071 León, Spain
{amigd,calvaa,acamv,vmato,agueh}@unileon.es

Abstract. Anomaly detection in network traffic is a crucial task for ensuring the security and integrity of communication systems. Traditional supervised machine learning models often achieve high accuracy but rely heavily on labeled datasets, which are costly to obtain and may become outdated. To address this limitation, this paper explores the use of unsupervised and online learning techniques for anomaly detection in network flow data. In this work, we compare three approaches: a baseline exact-match dictionary method, a supervised Decision Tree classifier, and an online One-Class SVM implemented using the River framework. The evaluation is performed on a real-world NetFlow-based dataset enriched with synthetic anomalies to simulate realistic threat scenarios. Results indicate that the online One-Class SVM achieves a high detection rate (recall = 0.9861) with a low false positive rate (FPR = 0.0118), highlighting its suitability for dynamic environments where adaptability and low maintenance are critical. This study demonstrates the potential of online unsupervised learning as a practical alternative to traditional models in network anomaly detection tasks.

Keywords: Anomaly detection · Network flows · Online learning · Unsupervised machine learning · Cybersecurity · One-Class SVM · NetFlow

1 Introduction

Anomaly detection in network flows is a cornerstone in the protection of critical infrastructures, as it enables the identification of unusual behaviors that may indicate attacks, misconfigurations, or unauthorized activities. In this context, flow-based traffic analysis, such as NetFlow [5] or IPFIX [1], has emerged as an effective strategy for monitoring large volumes of data without the need to capture the full content of network packets.

E. Corchado et al. (Eds.): CISIS 2025, CCIS 2807, pp. 13–22, 2026.
https://doi.org/10.1007/978-3-032-19770-2_2

The literature has addressed this challenge from multiple perspectives, with particular emphasis on the use of both supervised and unsupervised machine learning techniques. Supervised methods have shown high performance in detecting specific threats; however, their effectiveness relies on the availability of labeled datasets that accurately reflect the characteristics of the operational environment. This requirement is often difficult to satisfy in real-world scenarios, where labels may be scarce, outdated, or erroneous [7].

In light of these limitations, unsupervised approaches have gained prominence due to their ability to operate without labeled data during training, focusing instead on modeling normal behavior to detect significant deviations. Nevertheless, most of these approaches remain grounded in batch learning schemes, which present notable drawbacks in dynamic environments-such as the inability to adapt to concept drift and the need to store large volumes of data in memory.

With the rise of real-time systems, such as IoT devices, industrial networks, and distributed corporate environments, the paradigm of *online machine learning* has emerged as a promising alternative. This approach enables the incremental updating of models as new observations become available, thereby enhancing adaptability, reducing memory requirements, and improving integration in resource-constrained environments [16].

In this context, several prior studies have contributed to the understanding and development of anomaly detection strategies. Chandola et al. [4] offer a survey of anomaly detection techniques, highlighting the applicability of statistical, distance-based, density-based, and machine learning approaches across various domains, including network traffic. Within the realm of unsupervised learning, Kabir et al. [8] employed k-means clustering, while Schueller et al. [15] and Verkerken et al. [17] explored support vector machines (SVM) for flow-based anomaly detection. Wang et al. [18] introduced a hybrid method that combines behavioral profiling with network graph modeling to enhance detection accuracy.

The need for adaptability in non-stationary environments has motivated a growing body of work on online learning strategies. In a recent comparative study, Shahraki et al. [16] evaluated several online algorithms for detecting anomalies in network flows, highlighting their effectiveness in handling evolving patterns and concept drift. These contributions underscore the importance of moving beyond static models and adopting online approaches tailored to continuous and large-scale network monitoring scenarios.

This study proposes a network flow anomaly detection approach. By implementing and evaluating three different models-an exact-match-based method, a supervised decision tree, and a One-Class SVM adapted for *online learning*-we compare various approaches to assess their feasibility and effectiveness. The aim is to identify anomalous flows in a realistic environment using a dataset provided by a specialized software company, which has been augmented with synthetic anomalies that simulate unusual behaviors across different time periods and destination addresses.

The remainder of this paper is organized as follows: Sect. 2 details the methodology. Section 3 introduces the metrics used to assess performance, while

Sect. 4 presents and compares the results obtained. Section 5 discusses the findings and their implications, and finally, Sect. 6 summarizes the main contributions and outlines directions for future work.

2 Methodology

This section introduces the key components used in the proposed anomaly detection system. It describes the online learning framework, the dataset characteristics, and the detection algorithms that will later be integrated and evaluated.

2.1 Online Machine Learning

Before delving into the fundamentals of online learning, it is useful to briefly contextualize traditional batch learning methods. This strategy is characterized by a clear separation between training and validation phases, both of which require access to the entire dataset from the outset. However, this approach presents several drawbacks, including high memory consumption, susceptibility to concept drift, and the inability to adapt to previously unseen attributes [2].

In contrast, online learning represents a paradigm in which the model is updated continuously as new observations are received, without the need to store the entire historical dataset [16]. This methodology enables faster training with lower computational cost. Two of its main advantages are: (i) its feasibility on resource-constrained devices, such as embedded systems or sensors [6], and (ii) its ability to rapidly adapt to new data without requiring a full retraining process.

This approach proves especially useful in environments where data is generated continuously, as is the case in large-scale information analysis systems, anomaly detection, autonomous robotics, or humanmachine interfaces [6].

As in conventional learning, online learning can be categorized into supervised, unsupervised, and semi-supervised variants, depending on the availability of labels during the learning process [16]. In this study, an unsupervised approach has been adopted due to the challenges associated with obtaining labeled data in cybersecurity contexts. The manual annotation of network flows requires expert intervention and does not always ensure coverage of emerging threats, potentially leading to quickly outdated datasets.

In online unsupervised learning, data $D = (x_1, x_2, \ldots, x_m)$ are observed sequentially without labels. The objective is to construct a model $F \approx p(y \mid X)$ capable of detecting patterns or anomalies as new instances x_t are received. Unlike the batch learning paradigm, here the model is updated incrementally: at each time step t, a new version F_t is generated based on the current observation x_t and the previous state F_{t-1}, thus enabling continuous adaptation to a changing environment without the need to retain large volumes of data in memory.

2.2 Dataset and Preprocessing

The dataset employed in this study was provided by the company Proactivanet [14], consisting of network flows generated by internal company devices between July 8 and July 15, 2024. The data structure adheres to the NetFlow standard [5], with certain custom extensions.

Dataset Structure and Labeling. A notable addition is the `END_TYPE` attribute, indicating the termination reason of each flow: 1 = inactive timeout, 2 = normal termination, 3 = TCP flags.

Fields such as source and destination IPs may include IPv4, IPv6, or domain names. For privacy reasons, all addresses were anonymized using numerical encodings.

To evaluate detection performance, a total of 12,500 synthetic anomalies were programmatically inserted across the temporal window of the dataset. These were evenly distributed across five types (2,500 samples each):

- Benign IP to anomalous IP during business hours
- Benign IP to benign IP outside business hours
- Benign IP to anomalous domain during business hours
- Benign IP to anomalous domain outside business hours
- Anomalous IP to anomalous IP during business hours

The final dataset consists of 221,063 benign flows and 12,500 anomalous flows.

Preprocessing Transformations. Prior to model training and evaluation, the following preprocessing steps were uniformly applied:

- **Timestamp simplification:** Reduced to hour granularity to capture diurnal trends while mitigating temporal noise.
- **Port unification:** A new attribute `PORT` was created: source port for outgoing flows, destination port for incoming.
- **Byte count rounding (supervised models only):** For the Exact Match and Decision Tree classifiers, the number of bytes sent and received was rounded to the nearest multiple of 100. This transformation aims to reduce sensitivity to minor fluctuations in traffic volume and mitigate overfitting to exact values.

These transformations reduce feature sparsity and improve generalization, particularly for the exact match method, which is highly sensitive to minor variations.

It is important to note that anomalies in this context do not necessarily indicate malicious attacks. Rather, they represent behavioral deviations from typical patterns, such as unusual communication timings or uncommon destinations.

2.3 Algorithms

This section presents all the models that have been used in this article to detect network flow anomalies.

Exact Match Detector

The proposed method, referred to as the *Exact Match Detector*, is based on the construction of a dictionary for anomaly detection in network flows where the key is the flow characteristics. Initially, a model is "trained" using only benign flows. This model is subsequently used to identify anomalous flows in a test set, where any flow that differs even slightly from those observed during training is flagged as anomalous.

The features selected from the dataset to train the model include the source and destination IP addresses, ports, timestamp of the flow, direction (incoming or outgoing), and the number of bytes sent and received.

The purpose of the processing transformations described in Sect. 2.2 is to mitigate the effect of overly specific values, as the *Exact Match Detector* relies on exact matching to compare flows. Without such adjustments, trivial differences between training and test flows would lead to an increased false positive rate.

During the evaluation phase, each flow in the test set is assessed by searching for its corresponding key in the model's dictionary. If the key is not found, the flow is classified as anomalous. Otherwise, it is considered a normal flow.

Decision Tree

The *Decision Tree Classifier* is a supervised learning model that uses a tree-like structure to make decisions based on data attributes. Each internal node of the tree represents a condition on an attribute, each branch corresponds to the outcome of that condition, and each terminal leaf assigns a class label.

In the context of anomaly detection in network flows, the *Decision Tree Classifier* is constructed from a training set composed of labeled flows, including both normal and anomalous instances. During the training phase, the algorithm identifies the most relevant attributes for classification by evaluating criteria such as *entropy* or the *Gini index*.

In relation to preprocessing, the same procedure described previously has been applied, with additional encoding of IP addresses with the `LabelEncoder` method.

To optimize the model's performance, a grid search strategy was adopted using `GridSearchCV` from the Scikit-Learn library [13]. This procedure systematically explored combinations of hyperparameters to identify the configuration that yielded the best classification results on a validation subset.

One-Class SVM for Online Learning

For anomaly detection in network flows, a *One-Class SVM* model has been employed in its online learning variant using the River library [10]. This model is particularly suited for *novelty detection* tasks, as it enables the identification of deviations from normal behavior. Unlike the traditional batch approach, this variant updates the model incrementally, which is crucial in network environments characterized by real-time data and evolving patterns.

A warm-up phase is typically conducted prior to the first predictions to improve performance and enable a fair comparison with batch machine learning models [3]. Subsequently, as new flows are received, the model adjusts its incremental parameters without the need to retain past data, thereby mitigating the issues associated with storing large data volumes.

Regarding preprocessing, the same procedure described previously was followed. However, in this case, the data were scaled prior to model application. To achieve this, a reduced subset of flows was selected to train the scaler. Since the scaling algorithm operates in online mode, this initial warm-up phase is essential to stabilize its behavior and enable a progressive adaptation to the characteristics of the streaming data.

To optimize model performance, the parameter ν-which controls the expected proportion of anomalies in the dataset-has been tuned. Additionally, a learning rate scheduler based on `InverseScaling` has been configured, modulating the bias term according to the expression:

$$\eta_{t+1} = \frac{\eta}{(t+1)^p} \tag{1}$$

where p is a configurable parameter that was kept at its default value to ensure the progressive convergence of the model.

Since the River library does not natively support hyperparameter optimization, a custom script was developed to facilitate the configuration and experimentation process. This tool allows the specification of the parameters ν, η, and q, the selection of the online scaler, the number of warm-up flows, the subset of features to be used, and the strategy for IP address management.

The OCSVM model for online learning does not produce a direct binary classification; instead, it outputs a continuous score, where higher values indicate a greater likelihood of anomaly. However, since the score scale is unbounded, a quantile-based filtering mechanism has been implemented to determine a dynamic decision threshold. This threshold is adjusted according to a parameter q, which defines the quantile above which a flow is classified as anomalous. This technique enables continuous adaptation of the threshold, maintaining its effectiveness in the presence of changes in network traffic distribution.

3 Evaluation Metrics

In this work, the problem is a binary classification as it consists of identifying whether a network flow is benign or anomalous. The evaluation of the model is based on the confusion matrix, which comprises true positives (TP), true negatives (TN), false positives (FP), and false negatives (FN).

Based on these values, several metrics are defined to assess the model's performance. The **Accuracy** metric represents the proportion of correctly classified instances with respect to the total number of samples, and is computed as follows:

$$Accuracy = \frac{TP + TN}{TP + TN + FP + FN} \tag{2}$$

The **Recall** metric measures the model's ability to correctly identify anomalies, i.e., the proportion of true positives relative to the sum of true positives and false negatives. It is defined as:

$$Recall = \frac{TP}{TP + FN} \tag{3}$$

Finally, the **False Positive Rate (FPR)** indicates the proportion of false positives among all benign samples. It is calculated as:

$$FPR = \frac{FP}{FP + TN} \tag{4}$$

4 Results

This section presents the results obtained after implementing and evaluating the three models: *Exact Match Detector*, *DecisionTreeClassifier*, and *One-Class SVM*. Except for the Exact Match Detector, which has no tunable parameters, a hyperparameter optimization process was carried out for the other two models.

All models were trained using 100,000 network flows. In the case of both the Exact Match Detector and the One-Class SVM, all training flows were considered benign, meaning that no malicious instances were included during the training phase. However, as the *DecisionTreeClassifier* is a supervised model, a labeled dataset containing both benign and anomalous flows was required. Given the low prevalence of anomalies in the dataset, the model was evaluated using 6,250 anomalous flows and 6,250 benign flows.

The results for the *Exact Match Detector* reveal an **accuracy** of 0.741, a **recall** of 100%, and a **false positive rate** (FPR) of 0.518.

For the *DecisionTreeClassifier*, the optimal combination of hyperparameters included the use of the `entropy` criterion to evaluate split quality, no maximum depth (until all leaves are pure or contain fewer samples than `min_samples_split`), `min_samples_leaf = 1`, and `min_samples_split = 2`. This configuration achieved an **accuracy** of 0.9998, a **recall** of 0.9995, and a **FPR** of 0.0, indicating excellent predictive capacity with virtually no false positives.

Finally, the *One-Class SVM* model was optimized using the parameters $q = 0.99$, $\nu = 0.05$, and a learning rate $\eta = 0.25$. This configuration resulted in an **accuracy** of 0.9871, a **recall** of 0.9861, and a **FPR** of 0.0118.

Table 1 summarizes the performance results obtained for the three models using the dataset provided by the company Proactivanet.

5 Discussion

This section discusses the results obtained from the implementation and evaluation of the three models. The findings enable a comparative analysis of supervised and unsupervised approaches and allow for an assessment of each model's effectiveness in detecting anomalies in network flows.

Table 1. Performance of the three evaluated models using the dataset provided by Proactivanet.

Model	Evaluation Samples	Accuracy	Recall	FPR
Exact Match Detector	25,000	0.741	1.0	0.518
DecisionTreeClassifier	12,500	0.9998	0.9995	0.0
OSVM	25,000	0.9871	0.9861	0.0118

Firstly, the Exact Match Detector demonstrated a remarkable ability to detect all anomalies present in the dataset, achieving a recall of 100%. However, this high sensitivity came at the cost of a high false positive rate (FPR = 0.518), which compromises its practical applicability in operational environments where an excess of alerts may lead to operator fatigue and reduced trust in the system. Despite this limitation, the simplicity of the model makes it a viable option in scenarios where the priority is to maximize the detection of atypical events, regardless of the cost in false positives.

In contrast, the DecisionTreeClassifier exhibited exceptional performance, achieving an accuracy of 0.9998 and an FPR of 0.0. These results indicate that the model successfully distinguished between benign and anomalous classes, maintaining high precision while minimizing false positives. However, as a supervised model, its effectiveness may decline in scenarios with low anomaly prevalence or where attack patterns are not sufficiently represented in the training data. Furthermore, it requires a labeled dataset that is representative of the target network, which in real-world settings entails a substantial workload to obtain.

Finally, the One-Class SVM achieved a compelling balance between precision and false positive rate. Despite being an unsupervised model, it obtained a recall of 0.9861 and an FPR of 0.0118, suggesting that the tuning of the parameters ν, q, and η was appropriate to capture anomalous behaviors without generating an excessive number of false positives.

Overall, the results show that the DecisionTreeClassifier offers the best performance in terms of accuracy. However, its reliance on labeled data limits its applicability in environments where anomalies are infrequent or difficult to label. Therefore, the One-Class SVM emerges as the most viable option, as it operates in an unsupervised manner, making it more suitable for real-world scenarios [4]. It achieves a high detection rate (recall) and a low false positive rate (FPR), without requiring global retraining or the storage of large data volumes. Additionally, being an online model, it is capable of adapting to newly emerging anomalies-thereby addressing concept drift-by incrementally updating its internal representation as new data arrives. This property is essential in dynamic environments where traffic patterns evolve over time and static models rapidly become obsolete. Furthermore, due to its low computational footprint and unsupervised nature, the One-Class SVM can be integrated into existing IDS frameworks as an anomaly detection component, enhancing their ability to detect previously unseen or obfuscated threats.

While the dataset used in this study is based on real traffic and enriched with controlled synthetic anomalies, it is important to acknowledge that such anomalies may not capture the full spectrum of behaviors observed in complex threat scenarios. Nevertheless, this approach enables a reproducible and systematic evaluation of detection models under well-defined conditions, providing a solid foundation for further testing in more heterogeneous environments.

6 Conclusion

This work has comparatively evaluated three approaches to anomaly detection in network flows: an exact match method, a supervised decision tree, and an online One-Class SVM. While the supervised model (DecisionTreeClassifier) achieved the highest accuracy, its dependence on labeled data and limited scalability constrain its applicability. In contrast, the online One-Class SVM exhibited a favorable trade-off between detection capability and operational efficiency, offering a more adaptable and scalable solution for real-time network monitoring. The source code and dataset used in the experiments are publicly available in a GitHub repository [9].

As a future research direction, it would be interesting to explore hybrid architectures that combine online and offline models [11,12], leveraging the adaptability of the former and the statistical stability of the latter. Such a combination could further improve performance in dynamic environments, enhancing both the precision of detection and the efficiency of responses to anomalous events.

Acknowledgments. This research is a result of the CIBERLAB project (C083/23), carried out under the collaboration agreement between INCIBE and the University of León. This initiative is part of the Recovery, Transformation and Resilience Plan, funded by the European Union (Next Generation EU).

Disclosure of Interests. The authors have no competing interests to declare that are relevant to the content of this article.

References

1. Aitken, P., Claise, B., Trammell, B.: Rfc 7011: Specification of the ip flow information export (ipfix) protocol for the exchange of flow information (2013)
2. Bartz-Beielstein, T.: Introduction: from batch to online machine learning. In: Bartz, E., Bartz-Beielstein, T. (eds.) Online Machine Learning, pp. 1–11. Springer, Singapore (2024). https://doi.org/10.1007/978-981-99-7007-0_1
3. Bartz-Beielstein, T., Hans, L.: An experimental comparison of batch and online machine learning algorithms. In: Bartz, E., Bartz-Beielstein, T. (eds.) Online Machine Learning, pp. 105–124. Springer, Singapore (2024). https://doi.org/10.1007/978-981-99-7007-0_9

4. Chandola, V., Banerjee, A., Kumar, V.: Anomaly detection: a survey. ACM Comput. Surv. 41(3) (2009). https://doi.org/10.1145/1541880.1541882
5. Claise, B.: Cisco Systems NetFlow Services Export Version 9, issue: 3954 Num Pages: 33 Series: Request for Comments Published: RFC 3954 (2004)
6. Gepperth, A., Hammer, B.: Incremental learning algorithms and applications. In: European Symposium on Artificial Neural Networks (ESANN), Bruges, Belgium (2016)
7. Guerra, J.L., Catania, C., Veas, E.: Datasets are not enough: challenges in labeling network traffic. Comput. Secur. **120**, 102810 (2022)
8. Kabir, M.A., Luo, X.: Unsupervised learning for network flow based anomaly detection in the era of deep learning. In: 2020 IEEE Sixth International Conference on Big Data Computing Service and Applications (BigDataService), pp. 165–168. IEEE, Oxford (2020). https://doi.org/10.1109/BigDataService49289.2020.00032
9. Miguel-Diez, A.: amigueldiez/cisis25-anomaly-detection-OML (2025). https://github.com/amigueldiez/cisis25-anomaly-detection-OML
10. Montiel, J., et al.: River: machine learning for streaming data in Python (2021)
11. Odiathevar, M., Seah, W.K.G., Frean, M.: A hybrid online offline system for network anomaly detection. In: 2019 28th International Conference on Computer Communication and Networks (ICCCN), pp. 1–9. IEEE, Valencia (2019)
12. Odiathevar, M., Seah, W.K., Frean, M., Valera, A.: an online offline framework for anomaly scoring and detecting new traffic in network streams. IEEE Trans. Knowl. Data Eng. **34**(11), 5166–5181 (2022). https://doi.org/10.1109/TKDE.2021.3050400
13. Pedregosa, F., et al.: Scikit-learn: machine learning in python. J. Mach. Learn. Res. **12**, 2825–2830 (2011)
14. Proactivanet: Proactivanet: Software especializado en ITAM + ITSM (2025). https://www.proactivanet.com/
15. Schueller, Q., Basu, K., Younas, M., Patel, M., Ball, F.: A hierarchical intrusion detection system using support vector machine for sdn network in cloud data center. In: 2018 28th International Telecommunication Networks and Applications Conference (ITNAC), pp. 1–6. IEEE, Sydney (2018). https://doi.org/10.1109/ATNAC.2018.8615255
16. Shahraki, A., Abbasi, M., Taherkordi, A., Jurcut, A.D.: A comparative study on online machine learning techniques for network traffic streams analysis. Comput. Netw. **207**, 108836 (2022). https://doi.org/10.1016/j.comnet.2022.108836
17. Verkerken, M., D'hooge, L., Wauters, T., Volckaert, B., De Turck, F.: Unsupervised machine learning techniques for network intrusion detection on modern data. In: 2020 4th Cyber Security in Networking Conference (CSNet), pp. 1–8. IEEE, Lausanne (2020). https://doi.org/10.1109/CSNet50428.2020.9265461
18. Wang, W., Shang, Y., He, Y., Li, Y., Liu, J.: BotMark: automated botnet detection with hybrid analysis of flow-based and graph-based traffic behaviors. Inf. Sci. **511**, 284–296 (2020). https://doi.org/10.1016/j.ins.2019.09.024

Enhancing Malware Detection In Portable Executables with Random Forest and Variance-Based Feature Selection

Alexandru Todea(✉), Ciprian Pungilă, and Adrian Spătaru

Faculty of Informatics, West University of Timișoara, Timișoara, Romania
{alexandru.todea01,ciprian.pungila,florin.spataru}@e-uvt.ro

Abstract. This paper aims to propose an improved method for enhancing malware detection in portable executables (PE), through the Random Forest and variance-based feature selection mechanisms, in order to achieve higher accuracy than previous models, while decreasing training time. Our method involves training a Random Forest model with default parameters, its evaluation on the EMBER2018 portable executables dataset, and selection of relevant columns with variance exceeding a proposed threshold. We present our method's results, compare them with those of previous works done in the literature, and outline our approach's empirical benefits and reduced training time. Finally, we discuss potential ideas for future work and further improvements.

Keywords: Malware detection · Machine learning · Random forest · Feature selection · Portable executable

1 Introduction

In this paper, we tackle the binary classification problem of labeling Windows portable executable (PE) files as either malicious software (malware) or benign software. We define malware as a program that can harm the user. Malicious software includes, but is not limited to, viruses, worms, trojan horses, ransomware, spyware, adware, and rootkits.

Getting infected with the malware listed above can have serious consequences on individuals and corporations, including downtime, reputation loss, and disclosure of confidential information. More importantly, it can damage society's critical infrastructure. Detecting malware in a fast and accurate manner has become increasingly important in today's world to prevent its negative effects. To address this demand, we aim to develop a malware detection machine learning (ML) model that labels Windows PE files with a better accuracy than the current state of the art while maintaining training and inference times that are comparable or even shorter.

We use EMBER2018 [1] as a benchmark dataset because it is large (1.1 million samples) and includes a rich feature set. Many malware detection datasets

E. Corchado et al. (Eds.): CISIS 2025, CCIS 2807, pp. 23–32, 2026.
https://doi.org/10.1007/978-3-032-19770-2_3

are publicly available on platforms such as Kaggle [2], but they are small and provide a limited number of features. The only dataset that comes close in quality to the EMBER2018 dataset is SOREL-20M [3], but its large size (20 million samples) makes training models on it with a consumer-grade computer very difficult.

Section 2 summarizes the papers that performed experiments on the full EMBER2018 dataset. Section 3 details our experimental framework and the reasoning behind it. Section 4 presents the results of our experiments. Section 5 summarizes our work, integrates our findings into the wider body of research on the EMBER2018 dataset, and sets the stage for future studies.

2 Literature Review

2.1 Filtering the Literature

We searched Google Scholar for papers performing experiments on the EMBER2018 [1] dataset using two keywords: "ember" and "dataset". We applied a filter to only display papers published between 2018 and 2025.

We found that several do not train and evaluate the models on the full dataset. Those are not relevant to our study and will not be included in this literature review. There are roughly three types of papers that perform experiments on the full dataset: dataset feature analysis (FA) papers, traditional ML papers, and deep learning (DL) papers.

2.2 Dataset Feature Analysis (FA) Papers

The first category of papers analyzed focuses on FA. These papers try combinations between the feature groups in the dataset in an attempt to discover the ones that provide the best performance. The representative papers for this category were written by Oyama et al. [4] and Şandor et al. [5].

Oyama et al. [4] trained LGBM models on all possible feature combinations and evaluated their performance by computing scores using a formula defined by them. The formula contained an accuracy score parameter preceded by an accuracy weight coefficient. With the accuracy weight coefficient set to a value that placed four times higher importance on the accuracy parameter compared to the other parameters, the feature group combination that obtained the highest score was: general file information, header information, section information, and string information.

Şandor et al. [5] also considered all feature group combinations but tested the performance of models trained on them on four subsets with training and testing data extracted from the full EMBER2018 dataset. Before training the models, normalization, scaling, or no pre-processing was applied. Two types of models were trained and evaluated: Random Forest (RF) and logistic regression (LR). The RF models are more relevant to our study. Although experiments were performed on all subsets, the paper only reports Area Under the Receiver Operating Characteristic Curve (AUC) results on the largest subset at two false positive rate (FPR) thresholds: FPR less than 0.1% and FPR less than 0.01%. At FPR less than 0.1%, many RF models with no pre-processing and scaling pre-processing applied obtained the highest AUC score of 0.996. At FPR less than 0.01%, again, the RF models with no pre-processing and scaling pre-processing won. Multiple of them achieved the highest AUC score of 0.984.

2.3 Traditional Machine Learning (ML) Papers

The second category of papers evaluated included works that train and evaluate traditional ML models on the dataset. The representative papers for this category were written by Ghourabi [6] and Shashank et al. [7].

Ghourabi [6] proposed a security system for the healthcare industry that encompasses a malware detection system to protect the computers of medical staff against threats. The malware detection system uses a Bayesian-optimized LGBM model to label executables on the medical staff computers. The model was trained and evaluated on the full EMBER2018 dataset. It obtained an accuracy of 97.96%, outperforming an unoptimized LGBM and two neural networks (NN).

Shashank et al. [7] trained six ensemble ML models on the full EMBER2018 dataset. Normalization was performed to bring all columns to a mean of zero and a variance of one. A gradient-boosting-based feature reduction technique was used to select the most important features. RF, AdaBoost, Extra Trees, XGBoost, LGBM, and Bagging models were trained on the feature-reduced dataset. The best accuracy of 96.56% was obtained by the Bagging model.

2.4 Deep Learning (DL) Papers

The third category of papers studied included those that train and evaluate DL models on the dataset. The representative papers in this category were written by Connors and Sarkar [8] and Lad and Adamuthe [9].

Connors and Sarkar [8] trained and evaluated several ML models using the full EMBER2018 feature set and subsets of it. They obtained the best accuracy of 95.22% on the full feature set with a DL model designed by them. They are the only authors out of the bunch that we reviewed who made their code public. As a result, we successfully reproduced their experiment in Subsect. 3.2.

Lad and Adamuthe [9] used Scikit-learn's [10] StandardScaler to scale the full vectorized training and test data. Scaling prevents ML models from giving more importance to features with higher values [11]. The authors trained an NN that they created on the full EMBER2018 dataset and obtained an accuracy of 94.09%.

3 Methodology

3.1 Exploring and Pre-Processing the EMBER2018 Dataset

EMBER2018 is a large dataset that was created for malware detection research, containing an equal number of malicious and benign samples in both the training and testing sets. Table 1 shows the distribution of malicious, benign, and unlabeled samples in the dataset. We trained and evaluated several ML models on the EMBER2018 [1] dataset with the second feature version downloaded from [12].

Table 1. Distribution of benign, malicious, and unlabeled samples in the dataset

Dataset Partition	Benign	Malicious	Unlabeled
Training Set	300,000	300,000	300,000
Test Set	100,000	100,000	0
Total	400,000	400,000	300,000

The dataset contains 2,381 features extracted from the samples grouped into nine feature groups. The nine feature groups are: general file information, header information, imported functions, exported functions, section information, byte histogram, byte-entropy histogram, string information, and data directories. Detailed descriptions can be found in [13].

The dataset comes in the form of several large *.jsonl* files and needs to be vectorized (i.e., converted into numerical format) before ML models can be trained on it. We used the code provided by the dataset authors in [12] to perform vectorization. The instructions were not very detailed. As a result, we created a public GitHub repository [13] with clear steps on how to vectorize the dataset and run our experiments on the resulting vectorized version.

Anderson and Roth [1] included unlabeled samples in the dataset to encourage researchers to explore unsupervised and semi-supervised learning. Since we are only interested in supervised learning (i.e., learning with labels), we did not consider the unlabeled samples from the training set.

In Subsect. 3.2, we discuss how we reproduced the NN experiment performed by Connors and Sarkar [8]. However, before that, we took one more dataset preprocessing step that they took, namely, we shuffled the samples in the training set. To get the samples shuffled in the same order as them, we set the seed for NumPy's [14] random number generator to `314` (exactly as they did in the `NeuralNetwork.py` file in [15]) and then shuffled the training set. We also kept the samples in the shuffled order for the experiments in Subsect. 3.3.

3.2 Reproducing the Connors and Sarkar [8] Neural Network Experiment

We extensively searched for authors who made the code for their EMBER2018 experiments public, and the only ones we found were Connors and Sarkar [8]. The code for their experiments is available in [15]. Several papers from Sect. 2 failed to thoroughly document their experiments, which introduces ambiguity and impedes reproducibility.

Connors and Sarkar [8] trained and evaluated several ML models using both the full EMBER2018 feature set as well as reduced versions of it. However, we only reproduced their best-performing model, an NN trained on the full feature set, because that is the one we had to outperform for our work to be considered meaningful.

We followed the same training process as in [15]. We trained the NN model for 75 epochs, with a batch size of 200, on the 600,000 labeled samples from the dataset's training set and evaluated it on the 200,000 labeled samples from the dataset's testing set. No cross-validation was performed. We configured Keras [16] to save the model weights after the epoch with the highest accuracy on the testing set. We measured the training time. Once training was complete, we loaded the weights and computed the following performance metrics on the testing set: accuracy, number of true negatives (TN), number of false positives (FP), number of false negatives (FN), number of true positives (TP), AUC, precision, recall, F1 score (F1), and Cohen's Kappa Score. Additionally, we measured the inference time (i.e., the time it took the model to make the predictions on all the samples in the testing set).

3.3 Our Experimental Procedures

The first action we took was to perform a variance-based feature selection on the full EMBER2018 feature set. We used the Scikit-learn [10] VarianceThreshold to drop features with a variance lower than 0.001 from the dataset. 1,658 of the original 2,381 features remained. This decision was motivated by the book: Machine Learning for Tabular Data [17], in which Ryan and Massaron talk about certain conditions that should always be avoided among a dataset's features. One of them is "constant or quasi-constant columns". The authors mention that "the variance shouldn't approximate zero for numeric features", which is what we are dealing with in the case of the vectorized EMBER2018 dataset. They go on to explain that ML models can only learn from how the features vary with respect to the target and that "no change in the features implies no conditional change on your target from which to learn".

We trained and evaluated a total of four models: two Scikit-learn [10] RF classifiers and two LGBM classifiers [18]. One of the two RF models was trained using the variance-filtered (VF) EMBER2018 feature set, while the other was trained on the full EMBER2018 feature set. We proceeded in the same way for the LGBM models. All four models were trained and evaluated on the same training and testing sets as the NN detailed in Subsect. 3.2. No cross-validation was performed. The default parameters were used for all four models. We selected these models because we know from [17] that they provide good performance on tabular data while not consuming significant computational resources. The training and inference times were measured. The same performance metrics as for the NN were computed.

The best-performing model was the VF RF. To assess if its detection accuracy is stable across different testing sets, we computed a 95% confidence interval (CI) using bootstrapping. We created 5000 new testing sets of the same size as the original testing set by randomly drawing instances from it with replacement. We then computed the accuracy of the RF model, without changing its training set, on each of the 5000 testing sets. We sorted the results and identified the 2.5th and 97.5th percentiles to observe which values 95% of the accuracies fall between.

To show a statistically significant difference between the VF RF and the Connors and Sarkar [8] NN, we retrained and re-evaluated each model 10 times on the same training and testing sets as in the initial experiment. The accuracies from each iteration were recorded into separate lists for the RF and NN models. We then applied the Wilcoxon signed-rank test on the two lists and recorded the resulting p-value.

To the best of our knowledge, no other studies have explored individual feature selection on the EMBER2018 dataset. As discussed in Sect. 2, Oyama et al. [4] and Şandor et al. [5] have investigated feature group selection, but not the selection of individual features within the dataset's feature groups, which is what we accomplish in this study using a variance-based approach.

4 Results

All experiments were performed on a 2024 Apple Macbook Pro with an M4 MAX chip (16-core CPU, 40-core GPU) and 64 GB of unified memory [19].

Connors and Sarkar [8] report an accuracy of 95.22% for their NN model trained on the entire feature set. We were not able to obtain the exact same accuracy, but we came very close with an accuracy of 95.05% (see Table 2). Various factors, such as slightly different library versions or different randomization mechanisms, could have contributed to this small difference in results. Even if we were able to replicate the experiment exactly, our RF model trained on the VF dataset would still perform better, with an accuracy of 96.9% (see Table 2). That is a 1.85% increase in accuracy over the best-performing model trained by Connors and Sarkar [8], an NN trained on the entire feature set. As can be observed in Table 2, the 1.85% increase in accuracy translates to 3,693 more correctly labeled samples (2,592 false positives correctly reclassified as true negatives and 1,101 false negatives correctly reclassified as true positives). The AUC, which is reported in Table 3 and Fig. 1, is higher for the VF RF model compared to the Connors and Sarkar [8] NN, indicating superior classification performance.

The RF model trained on the VF dataset trained 7.1× times faster than the NN Model (3 min and 13.83 s vs. 22 min and 55.58 s). Its inference time for the 200,000 samples is 1.67× times faster than the NN Model (6.08 s vs. 3.64 s, see Fig. 2).

The CI in Fig. 3 indicates that 95% of the accuracies of the VF RF, computed across 5000 testing sets generated using bootstrapping, fall between 96.60% and 96.80%, with a mean of 96.70%. This narrow range suggests that the model's performance is stable and does not vary significantly across different testing sets.

The p-value obtained from the Wilcoxon signed-rank test was 0.001953, which is smaller than 0.05, meaning that there is a statistically significant difference between the two models. The accuracies for the VF RF ranged between 96.71% and 96.9%. The accuracies for the Connors and Sarkar [8] NN ranged between 94.69% and 95.19%.

Table 2. Confusion matrix metrics and classification accuracy across model variants

Model	TN	FP	FN	TP	Accuracy (%)
Connors and Sarkar [8] NN	95,289	4,711	5,184	94,816	95.05
VF RF	97,881	2,119	4,083	95,917	96.90
VF LGBM	91,211	8,789	5,097	94,903	93.06
RF (Full Dataset)	95,854	4,146	5,291	94,709	95.28
LGBM (Full Dataset)	92,342	7,658	4,490	95,510	93.93

Table 3. Comparative evaluation of model performance metrics on the test set

Model	AUC	Precision	Recall	F1	Kappa
Connors and Sarkar [8] NN	0.9851	0.9505	0.9505	0.9505	0.9011
VF RF	0.9949	0.9784	0.9592	0.9687	0.9380
VF LGBM	0.9833	0.9152	0.9490	0.9318	0.8611
RF (Full Dataset)	0.9899	0.9581	0.9471	0.9525	0.9056
LGBM (Full Dataset)	0.9858	0.9258	0.9551	0.9402	0.8785

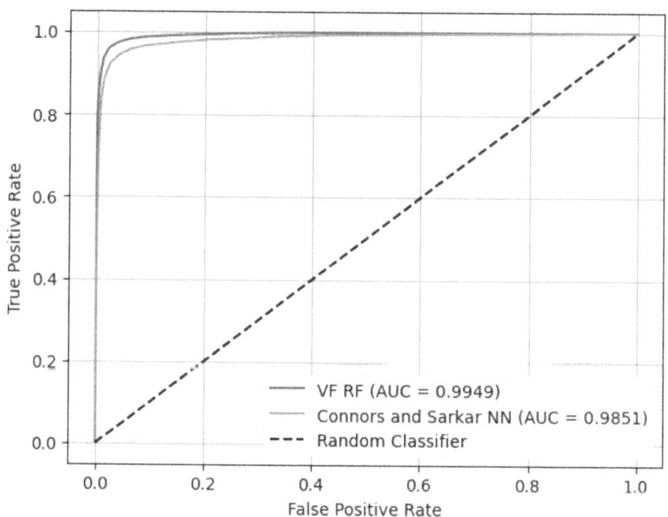

Fig. 1. ROC curves comparing the performance of the VF RF and the NN.

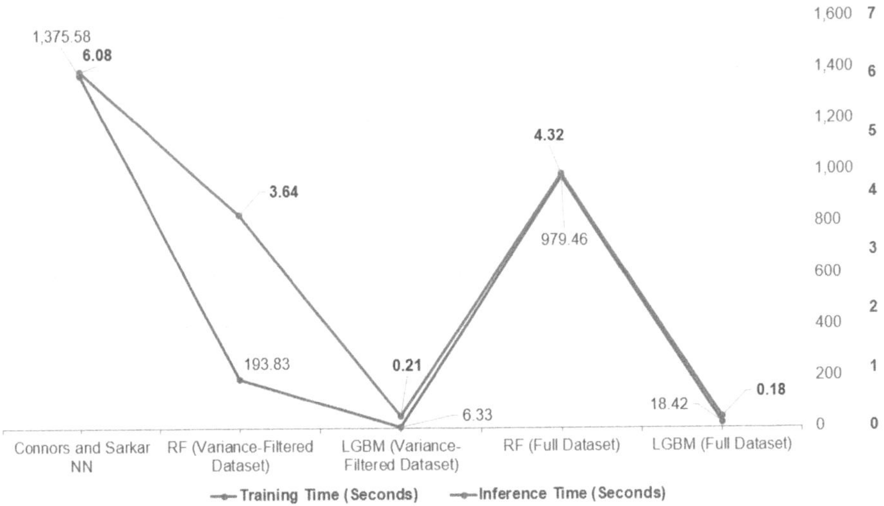

Fig. 2. Training duration and inference time for each model variant.

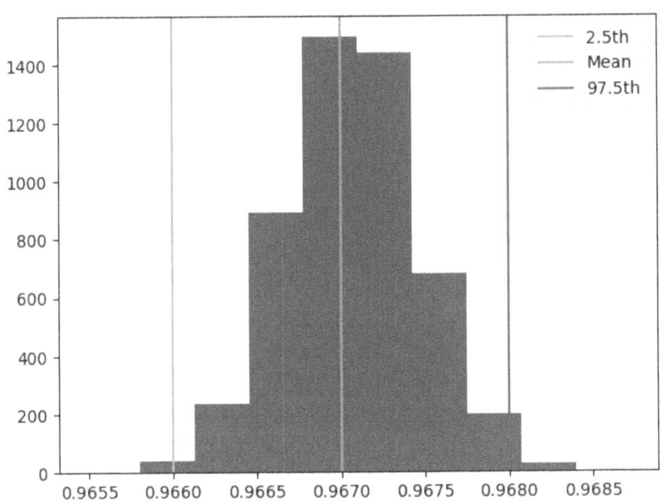

Fig. 3. 95% CI for the VF RF computed using bootstrapping with 5000 testing sets.

5 Conclusion

This paper has replicated the NN model trained on the full feature set of the EMBER2018 dataset [1] by Connors and Sarkar [8], and demonstrated that, through feature reduction and informed model selection, it can be outperformed in both accuracy and training time. Using our RF model trained and evaluated on a VF version of the EMBER2018 dataset, we achieved a 1.85% increase in accuracy while reducing training time by a factor of 7.1. The 1.85% increase in accuracy translates to 3,693 more correctly labeled samples, meaning that 3,693 more users would receive correct results for their PE file scans if the model were integrated in a real-life intrusion detection system (IDS). This, combined with the greatly reduced training time, which translates to faster updates to the known malware database, means that our approach would perform significantly better in the real world. The reduced update time is especially important given that hundreds of thousands of new malware are discovered daily [20].

Our accuracy is 2.81% better than the accuracy of 94.09% reported by Lad and Adamuthe [9] and 0.34% better than the accuracy of 96.56% reported by Shashank et al. [7]. Our accuracy is a little worse than the accuracy of 97.96% reported by Ghourabi [6]. However, it is not possible to verify if this is true, as the code or any implementation details for the Bayesian optimization were not provided. It is also not possible to compare training time with the papers that didn't make their code public.

Future work should focus on experimenting with more models that are known to behave well on tabular data. Multivariate feature selection could help improve the results if used alone or on top of univariate selection. Exploring embedded feature selection methods, possibly even in the context of RF, might lead to positive outcomes. If the resources allow, automated hyperparameter tuning and wrapper feature selection methods could be employed.

Aside from attempting to obtain better performance, future work can integrate our best model into an IDS. Two aspects must be considered when discussing such a system: processing new files inputted by the user and updating the system with new samples. Anderson and Roth [12] make these simple by providing code that takes as input a PE file and extracts from it the same features they extracted. It can be modified only to extract the individual features that remained after our variance filter. As a result, our model can make predictions on newly seen PE files. To update the system, the model needs to be retrained. After extracting the features with the code provided by Anderson and Roth [12], new samples can be added to the existing training set, and our model can be retrained.

Acknowledgments. This work has been partially supported by (1) the project RoNaQCI, part of EuroQCI, ID: DIGITAL-2021-QCI-01-DEPLOY-NATIONAL, 101091562, and (2) the project "Romanian Hub for Artificial Intelligence - HRIA", Smart Growth, Digitization and Financial Instruments Program, 2021-2027, MySMIS no. 351416.

References

1. Anderson, H.S., Roth, P.: EMBER: an open dataset for training static PE malware machine learning models, arXiv preprint arXiv:1804.04637 (2018)
2. Kaggle, Your Machine Learning and Data Science Community. https://www.kaggle.com. Accessed 10 Apr 2025

3. Harang, R., Filar, B.: SOREL-20M: a large scale benchmark dataset for malicious PE detection. In: Proceedings of the Virus Bulletin Conference (2020). https://github.com/sophos-ai/SOREL-20M. Accessed 10 Apr 2025
4. Oyama, Y., Miyashita, T., Kokubo, H.: Identifying useful features for malware detection in the EMBER dataset. In: Proceedings of the 2019 7th International Symposium on Computing and Networking Workshops (CANDARW), pp. 360–366. IEEE (2019)
5. Şandor, M., Portase, R.M., Coleşa, A.: EMBER feature dataset analysis for malware detection. In: Proceedings of the 2023 IEEE 19th International Conference on Intelligent Computer Communication and Processing (ICCP), pp. 203–210. IEEE (2023)
6. Ghourabi, A.: A security model based on LightGBM and transformer to protect healthcare systems from cyberattacks. IEEE Access **10**, 48890–48903 (2022)
7. Shashank, N.S., Madhu Kumar, S.D., et al.: Enhancing malware detection: a comparative analysis of ensemble learning approaches. In: Proceedings of the 2024 1st International Conference on Technological Innovations and Advance Computing (TIACOMP), pp. 35–40. IEEE (2024)
8. Connors, C., Sarkar, D.: Machine learning for detecting malware in PE files. In: Proceedings of the 2023 International Conference on Machine Learning and Applications (ICMLA), pp. 2194–2199. IEEE (2023)
9. Lad, S.S., Adamuthe, A.C.: Improved deep learning model for static PE files malware detection and classification. Int. J. Comput. Netw. Inf. Secur. **11**(2), 14 (2022)
10. The scikit-learn developers, Scikit-learn https://scikit-learn.org. Accessed 8 Apr 2025
11. Alshaher, H.: Studying the effects of feature scaling in machine learning, Ph.D. dissertation, North Carolina Agricultural and Technical State University (2021)
12. Anderson, H.S., Roth, P.: EMBER2018 dataset github repository. https://github.com/elastic/ember. Accessed 7 Apr 2025
13. Todea, A., Pungilă, C., Spătaru, A.: EMBER2018 experiments github repository. https://github.com/alexandrutodea/ember2018-experiments. Accessed 7 Apr 2025
14. The NumPy Developers, NumPy. https://numpy.org. Accessed 7 Apr 2025
15. Connors, C., Sarkar, D.: Machine learning for detecting malware in PE files – github repository. https://github.com/CollinConnors/Machine-learning-for-detecting-malware-in-pe-files. Accessed 7 Apr 2025
16. The Keras Developers, Keras. https://keras.io. Accessed 8 Apr 2025
17. Ryan, M., Massaron, L.: Machine Learning for Tabular Data, Manning Publications, February 2025. ISBN: 9781633438545
18. The LightGBM Developers, lightgbm.LGBMClassifier. https://lightgbm.readthedocs.io/en/latest/pythonapi/lightgbm.LGBMClassifier.html. Accessed 8 Apr 2025
19. Apple, MacBook Pro (16-inch, 2024) – Tech Specs. https://support.apple.com/en-us/121554. Accessed 9 Apr 2025
20. Estenssoro, J.V.: Malware and virus statistics 2025: the trends you need to know about https://www.avg.com/en/signal/malware-statistics. Accessed 23 June 2025

Relational Matrix of Cyberattacks: A Model for Threat Classification and Connection in Digital Environments

Mikel Ferrer Oliva(✉), José Amelio Medina Merodio, José Javier Martínez Herraiz, and Alberto Larena Luengo

Departamento de Ciencias de La Computación, Universidad de Alcalá, 28801 Madrid, Spain
{mikel.ferrer,josea.medina,josej.martinez,alberto.larena}@uah.es

Abstract. The increasing complexity of cyberattacks requires classification systems capable of representing how techniques interact and evolve. This paper introduces a relational matrix composed of eight attack groups and twenty directional connections that capture how one type of attack facilitates another. The matrix addresses limitations of static classifications by explicitly modeling interdependencies observed in documented incidents. Unlike hierarchical schemes, it enables bidirectional analysis of attack progressions and supports the incorporation of techniques without altering the overall structure. By identifying common escalation paths, the matrix enhances threat anticipation, improves strategic planning and supports early detection. Its structure offers a scalable and adaptable method for analysing adversarial behaviour across digital environments.

Keywords: Cybersecurity · relational matrix · threat classification · attack progression · dependency modeling

1 Introduction

Digital transformation has increased organisational exposure to cyber threats. Attacks have evolved from isolated events to coordinated sequences involving multiple techniques. This shift calls for classification systems that capture not only tactics but also their strategic interconnections. Efforts like MITRE ATT&CK [1], STIX 2.1 [2], and TAXII [3] have advanced threat systematisation and information sharing. Platforms such as MISP [4] enable categorisation using standard vocabularies. However, these models lack mechanisms to represent how attack phases are interdependent, limiting their capacity to identify enabling relationships [5]. Research shows that cyberattacks often follow multi-stage sequences. Social Engineering (SE) frequently initiates campaigns, particularly for credential compromise and Malware-Based Attacks (MBA) [6, 7]. Exploiting Software Vulnerabilities (ESV) also plays a key role in maintaining persistence [8, 9]. Still, current taxonomies do not adequately capture these facilitative chains or support forecasting across organisational layers.

Therefore, the objective of this work is to propose a relational matrix of cyberattacks that models the directional links between eight main attack groups. This structure enables

E. Corchado et al. (Eds.): CISIS 2025, CCIS 2807, pp. 33–42, 2026.
https://doi.org/10.1007/978-3-032-19770-2_4

a dynamic analysis of attack evolution and improves the representation of sequential risk scenarios. The paper is organized as follows. Section 2 presents the theoretical basis of the proposal and discusses the limitations of static models. Section 3 defines the attack categories and describes the relational matrix of connections. Section 4 discusses the implications of the model for incident analysis and strategic defense. Section 5 offers the main conclusions and outlines lines of future work.

2 Theoretical Foundation

The sophistication of cyber threats has led to models for classifying and analysing attacks. CASE [10] aimed to standardise event representation, but saw limited adoption. MISP [4] provides structured vocabularies for threat exchange, yet lacks the ability to model strategic dependencies between attacks. Many studies emphasise the need to understand how certain techniques enable later stages. Static taxonomies fail to reflect these interdependencies, reducing their value for anticipating escalation paths or analysing adversarial behaviour in complex scenarios [5].

2.1 Limitations of Static Models

Static taxonomies like MITRE ATT&CK [1] and MISP [4] offer structured listings of attack techniques but fail to reflect the dynamic interdependencies that define real-world incidents. By treating attacks as isolated events, these models lack the capacity to trace escalation paths or campaign progressions [5]. Studies show that cyberattacks typically unfold in stages: early vectors such as phishing often lead to credential theft and subsequent malware deployment [6, 7], while exploiting software vulnerabilities ensures persistence in later phases [8, 9].

Such transitions are rarely formalised, despite documented examples. Watering hole attacks compromise legitimate websites to deliver malware [11]; ransomware campaigns exploit remote access credentials [12]; and hybrid malware like *Lucifer* combines cryptojacking, exploiting software vulnerabilities (ESV) and DDoS to maximise disruption [13]. In IoT contexts, large-scale botnet attacks illustrate how device vulnerabilities enable chained intrusions [14, 15]. Capturing these sequential and tactical patterns is essential to model adversarial behaviour and anticipate threats. Without such modelling, static schemes remain disconnected from the needs of operational cybersecurity [16–18].

2.2 Consequences for Risk Management and Intelligence Sharing

The use of standardised formats like STIX 2.1 [2] and TAXII [3] facilitates consistent information exchange, yet they lack mechanisms to model sequential or enabling attack relationships. Their focus remains on isolated techniques, not on transitions between phases [16, 19]. Similarly, MISP [4] taxonomies support operational tasks but omit structural dependencies. CASE [10] aimed to unify ontologies but saw limited adoption and relevance. Studies show that seemingly isolated incidents often reflect coordinated strategies, where Social Engineering (SE) enables lateral movement and the exploitation of hidden vulnerabilities typically linked to Identity and Authentication Attacks (IAA) or

Exploiting Software Vulnerabilities (ESV) [9, 20, 21]. These chains are underrepresented in current standards, reducing predictive capability.

Complex threats like supply chain attacks demand models that capture coordinated tactics. Frameworks such as the Diamond Model [22] and VERIS [23] stress the need to map relationships to support strategic decisions and risk analysis [19].

3 Proposed Cyberattack Matrix

The relational model for classifying cyberattacks captures the interdependencies between adversarial actions. Unlike hierarchical or linear taxonomies, the matrix models dynamic links between attack groups, each defined by a strategic function and capable of acting as initiators or enablers within coordinated campaigns. It draws on documented incidents, technical reports, and scientific literature.

3.1 Classification of Attack Groups

The classification of attack groups is based on identifying both the initial vector and the adversary's strategic objective. Each group brings together techniques used in similar intrusion phases, facilitating the detection of operational patterns and improving defensive planning. Rather than clustering techniques by tools or technical similarities, the model adopts a functional logic that focuses on their tactical role within adversarial campaigns. Restricting the taxonomy to eight distinct groups ensures a balanced granularity: it avoids overlapping categories while capturing the most frequent and critical escalation paths. This structure enables clear mapping of entry vectors, transitions between phases and persistent threats, without fragmenting the model. Notably, complex attacks such as supply chain compromises—which often combine social engineering, malware deployment, software exploitation and long-term persistence—can be entirely described within this segmentation. Thus, the eight-group model preserves internal coherence while maximising explanatory coverage without requiring ad hoc categories or extensions. Table 1 summarises the resulting classification.

Table 1. Attack Groups Classification and Description.

Strike Group	Description
Social Engineering (SE)	Psychological manipulation to extract information or induce unsafe behaviour. Phishing is the most documented vector [6]
Malware-based attacks (MBA)	Malicious code to compromise systems, gain persistence or enable lateral movement (ransomware, trojans, botnets) [12]
Network Infrastructure Attacks (NIA)	Targeting routers and network devices to disrupt connectivity or integrity [19]
Exploiting Software Vulnerabilities (ESV)	Exploiting flaws to escalate privileges, execute code or access restricted resources [20]

(continued)

Table 1. (*continued*)

Strike Group	Description
Attacks on Protocols and Communications (APC)	Intercepting or manipulating communications (e.g. MitM, DNS hijacking) [24]
Identity and Authentication Attacks (IAA)	Credential-based attacks (brute force, stuffing) commonly used for initial access [7]
Attacks on critical IT/OT infrastructure (CIIA)	Actions against industrial systems (e.g. SCADA) often for sabotage or espionage [9]
APTs and Cyberespionage (APT)	Long-term intrusions aiming for stealthy data exfiltration or prolonged access [8]

3.2 Relational Dependencies Between Groups

The core of the matrix consists of twenty directional relationships between cyberattack groups, each reflecting typical sequences observed in real-world incidents. All relationships have been validated through documented cases, as detailed below:

IAA → SE: When attackers gain valid credentials, they can craft tailored social engineering strategies that enhance deception effectiveness [6, 7]. This is evident in SIM-swapping attacks, where adversaries tricked telecom providers into duplicating SIM cards, enabling interception of two-factor authentication codes [25].
IAA → MBA: Stolen credentials provide silent access that bypasses perimeter defences and enables malware deployment [12, 26]. This occurred in the DarkSide case, where attackers used compromised VPN credentials to access Colonial Pipeline's network and deploy ransomware that halted fuel distribution [27].
IAA → ESV: Privileged access allows discovery and exploitation of vulnerabilities that would otherwise remain inaccessible [15, 16]. In the Microsoft Exchange breach, attackers exploited ProxyLogon flaws after logging in with stolen credentials [28].
IAA → APC: Protocol-level authentication compromise enables interception or manipulation of communications without additional exploits [15, 20]. This occurred in the Target breach, where NTLM authentication was exploited using pass-the-hash techniques [29].
IAA → NIA: Administrative credentials allow attackers to reconfigure network devices and establish persistent access [14, 17]. This was demonstrated by the Mozi botnet, which compromised IoT routers using default passwords to launch peer-to-peer DDoS attacks [30].
SE → MBA: Phishing and other deceptive techniques often trick users into executing malware in trusted environments [6, 31]. In Emotet campaigns, users opened attachments that downloaded and executed malware at scale [32].
SE → ESV: Malspam can redirect victims to vulnerable applications, triggering remote code execution [11, 16]. Exploits like CVE-2017–11882 embedded in RTF documents enabled memory corruption in Microsoft Office [33].
MBA → APC: Certain malware variants are designed to interfere with communication protocols to spread or exfiltrate data [15, 20]. WannaCry exploited SMBv1 vulnerabilities to replicate across networks and disrupt file-sharing services [34].

MBA → NIA: Malware-infected devices are used to form botnets that degrade network integrity or launch large-scale attacks [9, 14]. Mirai used Telnet-accessible IoT devices to flood DNS providers and cause widespread outages [35].

MBA → CIIA: Malicious code can disable or manipulate critical control systems in industrial environments [9, 18]. NotPetya spread via a compromised software update and disrupted Ukraine's energy and logistics sectors [36].

MBA → ESV: Advanced malware often scans systems for unpatched vulnerabilities to escalate control [15, 16]. TrickBot loaders exploited EternalBlue flaws to gain deeper persistence in infected networks [37].

ESV → APC: Vulnerabilities in encrypted services allow attackers to intercept or extract confidential data [15, 20]. The Heartbleed flaw in OpenSSL enabled memory leaks from secure communication sessions [38].

ESV → NIA: Remote code execution enables lateral movement into core network components [15, 17]. Log4Shell enabled adversaries to breach internal infrastructure via exposed services [39].

ESV → APT: Advanced persistent threats exploit software flaws to maintain stealthy access in strategic systems [8, 16]. The SolarWinds attack inserted backdoors into the Orion platform using vulnerabilities affecting US government networks [40].

APC → NIA: Manipulating DNS and other protocols allows attackers to embed themselves in infrastructure and persist [14, 16]. In the Sea Turtle campaign, DNS records were altered to redirect authentication flows and compromise core systems [41].

APC → APT: Hijacked communication channels support long-term espionage and data theft [15, 16]. Sea Turtle attackers used DNS hijacking to harvest credentials and maintain covert access [42].

APC → CIIA: Protocol-level attacks in OT environments enable remote manipulation of industrial processes [9, 15]. Industroyer exploited IEC-104 to disable Ukrainian substations and trigger blackouts [43].

NIA → APT: Network compromises offer a base for long-term surveillance and pivoting into high-value systems [9, 18]. VPNFilter infected routers to intercept traffic and support advanced intrusion [44].

NIA → CIIA: Lateral movement from IT to OT systems can disrupt critical infrastructures [8, 18]. GreyEnergy breached corporate networks and reached industrial controls in the energy sector [45].

APT → CIIA: Long-term access by advanced threats enables sabotage or strategic data exfiltration [9, 16]. Stuxnet reprogrammed PLCs in Iranian nuclear facilities using multiple zero-days and forged certificates [46].

3.3 Cyberattack Relationship Matrix

To represent dependencies among attack groups, the proposed model adopts a relational matrix format. Each row corresponds to an initial vector, each column to a facilitated group, and binary values (1 or 0) indicate the presence or absence of direct relationships based on empirical evidence. This bidimensional structure enables a compact visualisation of connections and offers greater analytical flexibility than hierarchical models. For instance, Identity and Authentication Attacks (IAA) enable others such as

Social Engineering (SE), Malware-Based Attacks (MBA), or Exploiting Software Vulnerabilities (ESV), underscoring IAA's pivotal role in early intrusion phases. By modelling such links, the matrix helps identify critical escalation paths and design defence strategies based on how attacks structurally evolve. Importantly, the model includes 20 directed relationships, each selected through strict criteria of empirical traceability. This number reflects a deliberate focus on transitions that have been clearly and technically documented in real-world incidents (Table 2).

Table 2. Relational Matrix of Direct Cyberattack Dependencies

Initial Attack ↓/Facilitated Attack →	SE	MBA	ESV	APC	NIA	CIIA	APT
IAA	1	1	1	1	1	0	0
SE	0	1	1	0	0	0	0
MBA	0	0	1	1	1	1	0
ESV	0	0	0	1	1	0	1
APC	0	0	0	0	1	1	1
NIA	0	0	0	0	0	1	1

To support the interpretation of the relationships captured in the matrix, a visual framework was developed to represent the attack groups and the documented interconnections among them. This graphical representation is directly based on the proposed matrix structure, providing a concise overview of how transitions between different types of attacks are articulated (Fig. 1).

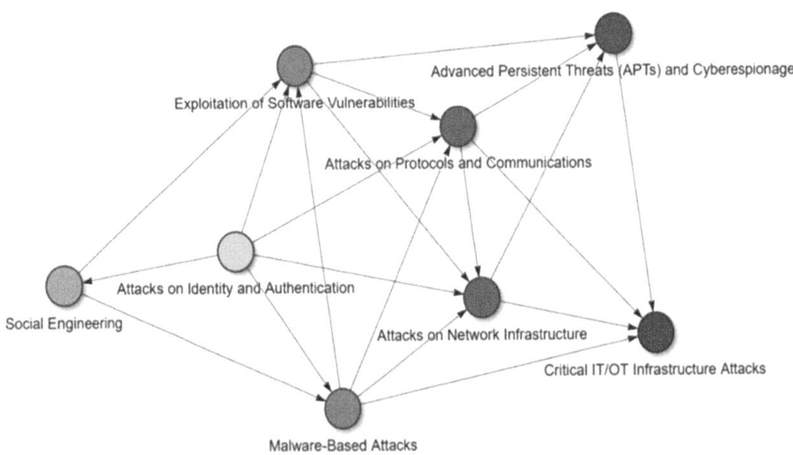

Fig. 1. Framework of attack groups and their interrelations.

4 Discussion

The relational classification of cyberattacks provides a structural response to the limitations of static and hierarchical models. Although traditional taxonomies organise techniques efficiently, they fail to capture how these interact within coordinated sequences [1, 4, 5], limiting their usefulness for forecasting and incident response. Cyber incidents often unfold through interdependent phases, with Social Engineering (SE) or Identity and Authentication Attacks (IAA) enabling steps such as Malware-Based Attacks (MBA), Exploiting Software Vulnerabilities (ESV) or Advanced Persistent Threats (APT) [6–9].

The proposed matrix bridges this gap by representing directional links between categories. Its twenty empirically validated relationships reflect real-world patterns observed in documented incidents and threat intelligence [15, 16, 20, 24], facilitating identification of escalation routes that static schemes often miss. Its relational structure allows new techniques to be integrated without altering the overall logic, ensuring scalability as threats evolve [5, 23]. It also identifies critical starting points: IAA links to five groups, highlighting its strategic relevance, while MBA's connections support its role in maintaining persistence and enabling lateral movement. Additionally, the model integrates the human factor. SE and IAA, shaped by user behaviour, function as entry points to more technical phases, emphasising the importance of behavioural control in cybersecurity programmes [21]. Finally, it aligns with models like the Diamond Model [22] and VERIS [23], but distinguishes itself by offering a compact, predictive classification framework suitable for both research and practice.

5 Conclusions

This work has presented a relational matrix for classifying cyberattacks, designed to represent the sequential and enabling nature of adversarial techniques. Unlike static models, it organises eight attack groups and twenty directional links, capturing how specific techniques facilitate others. By modelling these dependencies, the matrix improves the understanding of threat progression and supports the anticipation of complex escalation paths, addressing core limitations of conventional taxonomies.

At the organisational level, it enables early detection, optimises the allocation of defensive resources, and enhances incident response planning. It also identifies structurally critical attack groups that serve as early indicators of more advanced intrusions. From a social perspective, the model highlights the relevance of human factors by linking initial vectors to user behaviour patterns.

As a future line of research, this matrix may be further validated through empirical applications across various domains, including critical infrastructure and sector-specific environments. A key focus will be its alignment with structured threat representation languages. Specifically, the relational framework introduced here could be implemented in STIX 2.1 as custom objects or extended relationships that explicitly encode facilitation paths. These relationships may be shared through TAXII and integrated into MISP using standardised taxonomies, enabling interoperability and semantic precision in threat intelligence sharing.

Acknowledgments. This work has been developed within the "Recovery, Transformation and Resilience Plan", project C084/23 Ada Byron INCIBE-UAH, funded by the European Union (Next Generation).

References

1. ATT&CK: Adversarial Tactics, Techniques, and Common Knowledge, MITRE (2025). https://attack.mitre.org/. Accessed 6 June 2025
2. STIX Version 2.1. Committee Specification 02, O. C. T. I. TC (2025). https://docs.oasis-open.org/cti/stix/v2.1/cs02/stix-v2.1-cs02.html. Accessed 6 June 2025
3. TAXII Version 2.1. Committee Specification 01, O. C. T. I. TC (2025). https://docs.oasis-open.org/cti/taxii/v2.1/cs01/taxii-v2.1-cs01.html. Accessed 6 June 2025
4. CIRCL. MISP taxonomies and classification as machine tags. misp-project.org. https://www.misp-project.org/. Accessed 6 June 2025
5. ENISA. ENISA Threat Landscape for Supply Chain Attacks. European Union Agency for Cybersecurity (ENISA) (2021). https://www.enisa.europa.eu/sites/default/files/publications/ENISA%20Threat%20Landscape%20for%20Supply%20Chain%20Attacks.pdf. Accessed 6 June 2025
6. Bhardwaj, A., Sapra, V.: Why is phishing still successful?. Comput. Fraud Secur. 2020, 15–19 (2020). https://doi.org/10.1016/S1361-3723(20)30098-1
7. . Hellemann, D.N: Human risk review 2023. SoSafe Awareness GmbH, Colonia, Alemania, Report 2023. https://www.sosafe-awareness.com/. Accessed 6 June 2025
8. Connolly, L.Y., Wall, D.S.: The rise of crypto-ransomware in a changing cybercrime landscape: Taxonomising countermeasures. Comput. Secur. **87** (2019). https://doi.org/10.1016/j.cose.2019.101568
9. Clavijo Mesa, M.V., Patino-Rodriguez, C.E., Guevara Carazas, F.J.: Cybersecurity at sea: a literature review of cyber-attack impacts and defenses in maritime supply chains. Information **15**(11) (2024). https://doi.org/10.3390/info15110710
10. Community, C.: CASE: Cyber-investigation Analysis Standard Expression. CASE Community. https://caseontology.org/. Accessed 6 June 2025
11. Alrwais, S., et al.: Catching predators at watering holes: finding and understanding strategically compromised websites, pp. 153–166 (2016)
12. Beaman, C., Barkworth, A., Akande, T.D., Hakak, S., Khan, M.K.: Ransomware: recent advances, analysis, challenges and future research directions. Comput. Secur. **111**, 102490 (2021). https://doi.org/10.1016/j.cose.2021.102490
13. Networks, P.A.: Lucifer: new cryptojacking and DDoS hybrid malware exploiting high and critical vulnerabilities to infect windows devices. Palo Alto Networks - Unit 42, USA (2025). https://unit42.paloaltonetworks.com. Accessed 6 June 2025
14. Gelgi, M., Guan, Y., Arunachala, S., Samba Siva Rao, M., Dragoni, N.: Systematic literature review of IoT botnet DDOS attacks and evaluation of detection techniques. Sens. (Basel) **24**(11) (2024). https://doi.org/10.3390/s24113571
15. Wu, Q., Zhang, S., Zheng, B., You, C., Zhang, R.: Intelligent reflecting surface-aided wireless communications: a tutorial. IEEE Trans. Commun. **69**(5), 3313–3351 (2021). https://doi.org/10.1109/tcomm.2021.3051897
16. Rauf, U., Mohsen, F., Wei, Z.: A taxonomic classification of insider threats: existing techniques, future directions & recommendations. J. Cyber Secur. Mobil. (2023). https://doi.org/10.13052/jcsm2245-1439.1225

17. Salim, M.M., Rathore, S., Park, J.H.: Distributed denial of service attacks and its defenses in IoT: a survey. J. Supercomput. **76**(7), 5320–5363 (2019). https://doi.org/10.1007/s11227-019-02945-z
18. Niño, F.Y.Á.: Ransomware, una amenaza latente en Latinoamérica. InterSedes **24**(49) (2023). https://doi.org/10.15517/isucr.v24i49.50765
19. Hutchins, E.M., Cloppert, M.J., Amin, R.M.: Intelligence-driven computer network defense informed by analysis of adversary campaigns and intrusion kill chains. Lead. Issues Inf. Warfare Secur. Res. **1**(1), 80–106 (2011). https://www.lockheedmartin.com/content/dam/lockheed-martin/rms/documents/cyber/LM-White-Paper-Intel-Driven-Defense.pdf. Accessed 6 June 2025
20. Javeed, D., MohammedBadamasi, U., Ndubuisi, C.O., Soomro, F., Asif, M.: Man in the middle attacks: analysis, motivation and prevention. Int. J. Comput. Netw. Commun. Secur. **8**(7), 52–58 (2020). https://doi.org/10.13140/RG.2.2.22752.81928
21. Sedano Pinzón, J.J.: El contexto actual e histórico de la ingeniería social. LATAM Rev. Latinoamericana Ciencias Soc. Hum. **5**(5) (2024). https://doi.org/10.56712/latam.v5i5.2691
22. Caltagirone, S., Pendergast, A., Betz, C.: The Diamond model of intrusion analysis (2013). https://doi.org/10.13140/RG.2.2.31143.56481
23. Community, V.: VERIS: vocabulary for event recording and incident sharing. VERIS Community. https://verisframework.org/. Accessed 6 June 2025
24. CISA and FBI. TrickBot Malware. Cybersecurity and Infrastructure Security Agency (CISA), Federal Bureau of Investigation (FBI), USA, AA21–076A (2021). https://www.cisa.gov/news-events/alerts/2021/03/17/cisa-fbi-joint-advisory-trickbot-malware-0. Accessed 6 June 2025
25. Insikt Group. The Business of Fraud: SIM Swapping. Recorded Future (2021). https://go.recordedfuture.com/hubfs/reports/cta-2021-0825.pdf. Accessed 6 June 2025
26. Ivanov, M., Kliuchnikova, B., Chugunkov, I., Plaksina, A.: Phishing attacks and protection against them, pp. 425–428 (2021). https://doi.org/10.1109/ElConRus51938.2021.9396693
27. CISA and FBI, Darkside Ransomware: Best Practices for Preventing Business Disruption from Ransomware Attacks. Cybersecurity and Infrastructure Security Agency (2021). https://www.cisa.gov/sites/default/files/publications/AA21-131A_Darkside_Ransomware.pdf. Accessed 6 June 2025
28. CISA. Mitigate Microsoft exchange on-premises product vulnerabilities. Cybersecurity and Infrastructure Security Agency (CISA) (2021). https://www.cisa.gov/ED2102. Accessed 6 June 2025
29. United States Senate Committee on Commerce, Science, and Transportation, A "Kill Chain" Analysis of the 2013 Target Data Breach. United States Senate Committee on Commerce, Science, and Transportation (2014). https://www.commerce.senate.gov/services/files/24d3c229-4f2f-405d-b8db-a3a67f183883. Accessed 6 June 2025
30. E. Council to Secure the Digital, Ustelecom, and A. Consumer Technology. International botnet and IoT security guide 2021. USTelecom; Consumer Technology Association (CTA) (2021)
31. Álvarez, A.L., Cruz, J.A., Cruz, S.B., Gallardo, J.D.C., López, I.M., García, R.E.: El phishing como amenaza en la ciberseguridad corporativa de grandes empresas. Investigaciones Latinoamericanas en Ingeniería y Arquitectura (1), 26–33 (2024). https://doi.org/10.51378/ilia.vi1.8496
32. ANSSI. The Malware-As-Service-Emotet. Agence Nationale de la Sécurité des Systèmes d'Information (ANSSI) (2021). www.cert.ssi.gouv.fr
33. Cysiv. Threat Report: Formbook Infostealer. Cysiv Inc. (2021). www.cysiv.com. Accessed 6 June 2025
34. Smart, W.: Lessons learned review of the WannaCry Ransomware cyber attack. NHS England (2018). www.gov.uk/dh. Accessed 6 June 2025

35. Cloudflare. DNS and the Threat of DDoS. Cloudflare Inc. (2022)
36. Steinberg, S., Stepan, A., Neary, K., Rattray, G., Healey, J.: NotPetya: a columbia university case study. Case Consortium @ Columbia, School of International and Public Affairs, Columbia University, SIPA-21-022.1 (2021)
37. ANSSI. RYUK RANSOMWARE. Agence Nationale de la Sécurité des Systèmes d'Information (ANSSI) (2021). www.cert.ssi.gouv.fr
38. ICS-CERT. ICS-CERT MONITOR January - April 2014. U.S. Department of Homeland Security (DHS), National Cybersecurity and Communications Integration Center (NCCIC) (2014)
39. F. B. o. I. F. Cybersecurity and Infrastructure Security Agency (CISA), National Security Agency (NSA), Australian Cyber Security Centre (ACSC), Canadian Centre for Cyber Security (CCCS), Computer Emergency Response Team New Zealand (CERT NZ), New Zealand National Cyber Security Centre (NZ NCSC), United Kingdom's National Cyber and S. C. (NCSC-UK). Mitigating Log4Shell and Other Log4j-Related Vulnerabilities. CISA; FBI; NSA; ACSC; CCCS; CERT NZ; NZ NCSC; NCSC-UK, Product ID: AA21–356A (2021)
40. s. Federal Energy Regulatory Commission, I. Electricity, and C. Analysis Sharing. SolarWinds and Related Supply Chain Compromise: Lessons for the North American Electricity Industry. Federal Energy Regulatory Commission (FERC); Electricity Information and Analysis Sharing Center (E-ISAC) (2021)
41. CISCO. Threats of the Year: A look back at the tactics and tools of 2019. In: Cisco Cybersecurity Series 2019. Threat Report. Cisco (2019). http://www.cisco.com/go/securityreports. Accessed 6 June 2025
42. Cisco Talos. 2024 Cisco Talos Year in Review. Cisco (2025). https://blog.talosintelligence.com/content/files/2025/03/2024YiR-report.pdf. Accessed 6 June 2025
43. Cherepanov, A.: WIN32/INDUSTROYER a new threat for industrial control systems. ESET (2017)
44. Symantec. ISTR. Internet security threat report volume 24. Symantec (2019)
45. Cherepanov, A.: GREYENERGY a successor to BlackEnergy. ESET (2018)
46. Shakarian, P.: Stuxnet: Revolución de Ciberguerra en los Asuntos Militares. Air Space Power J. (2012)

Cryptography and Blockchain Security

Securing On-Chain Voting Using A Two-Layered zk-SNARK Approach

Gheorghe-Mădălin Pahomi(✉), Darius Galiș, and Ciprian Pungilă

Faculty of Informatics, Department of Digital Technologies and Software Engineering, West University of Timișoara, Bulevardul Vasile Pârvan 4, Timișoara 300223, Timiș, România
gheorghe.pahomi94@e-uvt.ro

Abstract. We present an architecture for double Zero-Knowledge Succinct Non-Interactive Argument of Knowledge (zk-SNARK) blockchain-based system for on-chain voting that aims to focus on privacy. The proposed architecture uses a two-stage approach, by creating a proof of eligibility for the voters, without revealing sensitive identity attributes, while enabling anonymous and verifiable vote submissions through the blockchain. This article explores the motivation for proposing the system, as well as describing the technical components of the system, together with an assessment of its technical features from a privacy and attack resiliency point of view. Experimental results show proof generation in under 2 s and verification in under 1.5 s on modest hardware. On-chain deployment costs approximately 0.001 SepETH, with each vote verification requiring only about 0.0001 SepETH, highlighting the system's efficiency and affordability for real-world use.

Keywords: On-Chain Voting · ZKP · Zero-Knowledge Proof · zk-SNARK · Blockchain

1 Introduction

Electronic voting systems, like traditional ones, must ensure integrity, verifiability, and secrecy. Zero-knowledge mechanisms offer a promising solution to meet these security needs. Blockchain-based voting enables remote, transparent, and private elections by combining smart contracts with zero-knowledge proofs. This paper focuses on maintaining transparency and voter anonymity through these technologies.

The concept of zero-knowledge proof (ZKP) was first mentioned in a study from 1989, being defined as "those proofs that convey no additional knowledge other than the correctness of the proposition in question" [1]. They allow one party (the prover) to convince another party (the verifier) that a statement is true, without revealing information about the statement itself. The same article also introduced the concept of interactive zero-knowledge proofs, which involve multiple rounds of interaction between the prover P and the verifier V. In contrast, non-interactive proofs involve the prover using a common reference string

E. Corchado et al. (Eds.): CISIS 2025, CCIS 2807, pp. 45–55, 2026.
https://doi.org/10.1007/978-3-032-19770-2_5

obtained from a trusted setup to obtain a mono-directional communication, from P to V [2]. There are two major directions that employ zero-knowledge proofs:

- `zk-SNARK`: One of the non-interactive types of zero-knowledge is the Zero-Knowledge Succinct Non-Interactive Argument of Knowledge (`zk-SNARK`). It generates short, fast-to-verify proofs without interaction between parties, by using a proving system based on elliptic curve pairings. It has applications in blockchain systems where privacy and efficiency are critical, and interactivity cannot be achieved as easily.
- `zk-STARK`: Zero-Knowledge Scalable Transparent Argument of Knowledge is another non-interactive type that is designed to provide transparency and scalability while maintaining cryptographic security [3]. It also has applications in blockchain protocols, to improve scalability, being suitable for complex computations of large-scale systems. Unlike `zk-SNARK`, which requires a setup phase, `zk-STARK` eliminates the need for a trusted setup, being more transparent and less prone to risks associated with compromised setups. It uses a proving system based on homomorphic encryption.

We believe that the Groth16 proving system [4] is favorable due to its combination of succinctness, fast verification, and integration with existing developer tools (for example, Circom [5]) which simplify backend and frontend circuit integration with low gas costs. `zk-STARK`, despite having the advantage of scalability and transparency, it generates larger, slower-to-verify proofs, making it less suitable for blockchain-based voting.

The present paper is structured as follows: Sect. 1 explores the motivation and presents the methods used in our system. Section 2 explores voting systems relevant to our paper. Section 3 details the architecture of our proof of work. Section 4 presents a taxonomic assessment of our system. Section 5 concludes the paper with a summary of our work.

2 State of the Art

To identify relevant electronic voting systems, we conducted a literature review using Google Scholar, IEEE Xplore, and the ACM Digital Library, focusing on recent publications and technical reports describing real-world implementations. Our search used keyword combinations like "open source electronic voting system," "secure voting system," "end-to-end verifiable voting," "blockchain voting systems," and "`zk-SNARK` voting systems." Without restricting the timeframe, we included systems with demonstrated long-term viability, identifying implementations dating back to 2005 that remain in use today.

Helios [6] is a web-based open-audit voting system that uses homomorphic ElGamal encryption, various protocols to prevent coercion and ensure anonymity between communicating parties, and proofs of decryption. Helios is not using blockchain, but a bulletin board (web server) where encrypted ballots are stored. Voter privacy is achieved by encrypting the votes with a public key, while the decryption keys are held by a set of trustees. The security aspect of Helios relies not on their server, but on the honesty of the parties that hold the decryption keys.

Civitas [7], is a coercion-resistant voting system based on research published in 2005 [8] that introduced dual credentials (one real, one fake) to prevent vote tampering. It extends this model with mix-nets and zero-knowledge proofs to ensure vote validity and anonymity. Like our approach, Civitas also considers election costs, though it does not use blockchain, relying instead on a public bulletin board for encrypted votes. Privacy is ensured through ElGamal encryption, mix-networks, and random permutations, proved with the help of zero-knowledge protocols. However, the focus on coercion-resistance hinders the scalability of Civitas by having intensive cryptographic operations and protocols that add complexity by sharing responsibilities among agents.

Semaphore [9] is an open source blockchain voting system that incorporates a Merkle tree of eligible voters and a `zk-SNARK` approach, with the proof being verified by a smart contract. Its design makes Semaphore more suitable for group-based anonymous signalling, but adds complexity on the user side, translating to technical difficulties for the general public, since the voter has to manage cryptographic operations and files. The high gas fees that result from proof verification should also be taken into consideration.

Although there are not many implementations, and even fewer documented ones, research in this direction shows clear interest in achieving a digital alternative to physical elections.

3 Methodology

Our architecture (see Fig. 1) is composed of four main stages, each corresponding to a distinct phase of the secure voting process.

- **1. Voter Authentication and Session Initialization.** The voter authenticates via a browser-based extension (i.e. MetaMask), establishing a secure session. A key exchange using Diffie-Hellman derives a shared secret, from which a session-specific AES-GCM encryption key is generated via HKDF. This key secures identity data (SSN and UID) even if HTTPS is unavailable. The backend generates a session nonce, signed via ECDSA, which is used to bind proof computation to the session.
- **2. `zk-SNARK-1`: Eligibility Verification.** The backend uses `zk-SNARK-1` to prove voter eligibility based on encrypted SSN and UID pairs received from the frontend, along with the nonce, after an initial eligibility check of the received pair against a stored SHA-256 database. The arithmetic circuit, implemented in Circom using R1CS constraints, ensures that only valid identity pairs can produce a valid proof. A cryptographic nullifier is generated to prevent double voting.
- **3. `zk-SNARK-2`: Vote Integrity Proof.** Upon vote casting, `zk-SNARK-2` is used to validate the vote. It consumes the nullifier from `zk-SNARK-1` and the voter's selection, producing a proof and a commitment. Only the nullifier and commitment are submitted on-chain, ensuring privacy and efficiency.

- **4. Blockchain and Smart Contract Integration.** The Solidity smart contract governs the on-chain logic: enforcing vote windows, preventing duplicate votes via nullifiers, and tallying votes. It does not validate proofs on-chain (for gas efficiency) but can be modified to do so if trust in the backend is compromised. A MetaMask-integrated frontend handles proof submission and vote visualization, including a verifier interface for public audits.

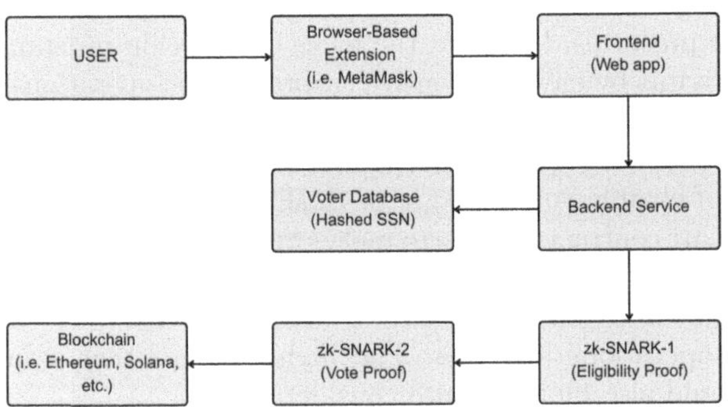

Fig. 1. Proposed architectural workflow.

Figure 1 presents the complete system architecture, detailing the four main stages: voter authentication, eligibility proof generation (`zk-SNARK-1`), vote commitment generation (`zk-SNARK-2`), and on-chain integration. It shows how each component contributes to privacy, auditability, and integrity.

The user is checked against a hashed database containing personally identifiable information (i.e. social security number, user ID, etc.), making sure the voter is legitimate. The backend uses `zk-SNARK-1` to generate a proof of eligibility and a nullifier. This ensures fair computation and prevents double voting. After a vote is cast, the backend calls for a second `zk-SNARK` (further referred to as `zk-SNARK-2`) that takes the public output of `zk-SNARK-1` (the nullifier) along with the voting choice and generates a proof that confirms the correctness of the computation and a commitment, which is sent to the blockchain. Since only the nullifier and vote commitment are submitted on-chain, we avoid the high cost and limitations of on-chain `zk-SNARK` verification, while maintaining the transparency benefits in blockchain technology. Computations are done cost-free off-chain, while the blockchain guarantees: a) *transparency* by publicly recording the votes so that anyone can audit that no nullifier is reused and each vote is valid; b) *integrity* by ensuring that once a vote is submitted, it cannot be altered, deleted, or tampered with, and c) *decentralization* by enforcing voting rules through a smart contract, not a central authority, namely no single entity has unilateral control over the election.

The voter database is a server-hashed list (using SHA-256 encryption protocol) of individuals that meet voting criteria (age, citizenship, mental and legal capabilities). The values used in the database are the social security number (SSN) and a User Identification (UID). The former one is a unique number (thirteen digits in Romania, for example) that links to an individual and is used in finding information from the government population evidence (making it possible to find if the voter is eligible). The upper limit of digits is given by the Groth16 system that uses BN254 elliptic curve, which operates over a 254-bit prime field. The circuit can also be adapted to use ASCII encoding, similarly to the UID. The latter can be any secret value that acts like a password, in order to make the authentication less susceptible to brute force attacks. It can either be user chosen (then only the SSN would be verified in the database to check eligibility), can be mandated by the government, or, as in this case, can be the national ID's batch and number. The credentials are never stored or transmitted in plaintext, while the proof is generated by the zero-knowledge mechanism.

`zk-SNARK-1` uses the SSN, the UID, and a randomly generated session nonce (used to prevent replay attacks and to bind the generated proof to a specific session). This nonce value is signed by an ECDSA to ensure that it actually came from the server, protecting it from various malicious attacks (e.g. client generating arbitrary nonces).

To implement `zk-SNARK-1`, we used Circom, a domain-specific language for constructing arithmetic circuits, which adheres to the Rank-1 Constraint System (R1CS) model. The circuit accepts as inputs the voter's Social Security Number (SSN), a User ID (UID) string, and a session-specific nonce. The UID is parsed as an array of ASCII-encoded characters, and we compute the cumulative product of these values, denoted as $p_i = \prod_{j=0}^{i} \texttt{uid}[j]$.

Using this, the internal variables of the circuit are derived as follows:

- $a = \texttt{SSN} \times p_i$
- $b = a + \texttt{nonce}$, encoded in R1CS form as $(a + \texttt{nonce}) \times 1 = b$
- $c = b \times b$

Where p_i represents the cumulative product of UID characters parsed as ASCII values, with uid[j] being the ASCII value of the j-th character in the UID array, `a` is the product of the SSN and p_i, binding both private identifiers, `b` is the sum of `a` and the session number, binding the computation to a specific session and `c` is the square of `b`, which increases entropy before the Poseidon hash is applied.

The output of this arithmetic chain is hashed using the Poseidon hash function to derive the final nullifier:

$$\text{nullifier} = \text{Poseidon}\,(\texttt{c})$$

This construction ensures that the nullifier uniquely ties the proof to the voter's identity and session, without revealing any of the raw inputs.

Figure 2 illustrates the R1CS representation of the `zk-SNARK-1` eligibility circuit. It highlights how SSN, UID (as ASCII values), and the session nonce are multiplied and combined into an arithmetic chain, whose final result is hashed using Poseidon to derive a unique voter nullifier.

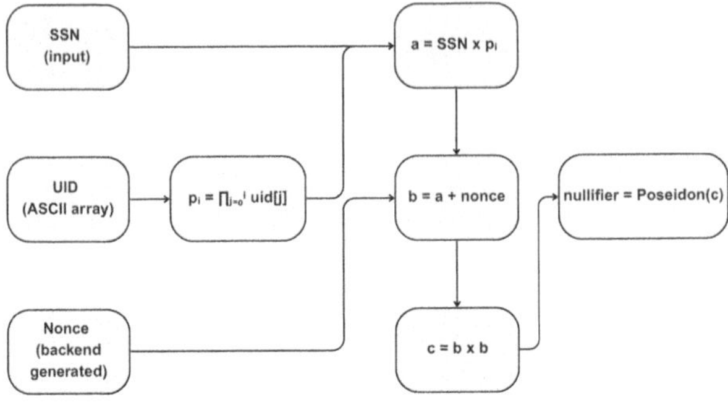

Fig. 2. Nullifier circuit.

The core of our system is the Poseidon hash function, a cryptographic hash optimized for efficiency and used by Circom. We employ Groth16, a `zk-SNARK` proving system that generates succinct, efficiently verifiable proofs for quadratic arithmetic programs (QAP) over elliptic curve points, derived via the Powers of Tau Ceremony [10]. After compiling the eligibility and vote logic into R1CS and then QAP, a trusted setup produces the proving and verification keys per circuit. The backend uses the proving keys to generate two `zk-SNARK` proofs, which are verified off-chain using the verification keys to ensure they match the circuit logic and public outputs. This guarantees voter eligibility and vote validity. Our eligibility R1CS includes 425 constraints, the vote R1CS 517, and we used a trusted setup supporting up to 32,768 constraints for scalability without redoing the ceremony.

We have tested our approach using a smart contract created in Solidity [11], in an Ethereum [12] based ecosystem. The contract governs on-chain logic by securely handling vote casting, preventing duplicate votes, enforcing voting deadlines, and recording results, without exposing voter identity or vote content. It accepts votes only within a fixed time window, stores the nullifier and commitment, and increments each candidate's individual vote counter, revealing the leading candidate only after the voting period ends.

To reduce gas fees, the contract accepts only valid proofs without performing on-chain verification, which would increase complexity and cost, a compromise that could be taken into consideration if the party submitting the proofs cannot be trusted. The smart-contract interaction is abstracted by a frontend, allowing interaction through MetaMask [13]. A verifier page that acts as an audit interface is also provided where users can upload `zk-SNARK` proof files, including their hashes, to validate Groth16 proofs and receive a match or mismatch result in the interface. To better understand the threat surface and how our system mitigates specific vulnerabilities, Table 1 summarizes the principal attack types and corresponding defense mechanisms across all layers of the architecture.

Table 1. Threat Model and Mitigation Strategies

Threat Type	Description	Mitigation Strategy
MITM Attacks	Attacker intercepts communication between client and backend	Use signed nonces, TLS/HTTPS, and Diffie-Hellman with HKDF
Frontend Tampering	Compromised JavaScript delivers malicious logic	Subresource Integrity (SRI), client-side verification interface
Replay Attacks	Reuse of previously valid messages (e.g., votes or credentials)	Session-specific ECDSA-signed nonces and unique nullifiers per voter
Sybil Attacks	Fake identities attempt to overwhelm the system	Use of a hashed registry and strict SSN:UID eligibility checks
Proof Forgery	Malicious clients submit fake `zk-SNARKs` or tampered proof files	Hash the proof files with SHA-256 before verification and accept only proofs verified using the correct key
Double Voting	A single user tries to vote more than once	Nullifiers submitted and tracked on-chain to enforce one vote per user
Server Malfunction	Backend incorrectly generates or verifies proof	Modular architecture allows off-chain auditing and detailed logging

4 Taxonomic Assessment

Electronic voting must balance ballot privacy with election integrity. Privacy protects voters from coercion or vote buying, while integrity ensures all votes are accurately recorded, tallied, and verifiable to prevent fraud or manipulation. With those characteristics in mind, we validated our approach against the ones presented in the state-of-the-art chapter.

To assess our approach against others, we performed a comparison using different criteria, splittable in two primary categories: **1. Exploitability metrics:** Attack vector (AV), Attack complexity (AC), Privileges required (PR), User interaction (UI), Scope (S) and **2. Impact metrics:** Confidentiality impact (C), Integrity impact (I), Availability impact (A) [14]

Table 2. Taxonomic evaluation of electronic voting systems

	AV	AC	PR	UI	S	C	I	A
Helios	Network	Low	None	Required	Unchanged	High	High	None
Civitas	Network	High	High	None	Unchanged	High	High	None
Semaphore	Network	High	Low	None	Unchanged	High	High	Low
Our proposed system	Network	High	None	None	Unchanged	High	High	None

The values for our system in Table 2 are derived from the structure and implementation of our voting architecture. The attack vector is "Network" due to web-based interactions through MetaMask. Attack complexity is rated "High" because of the layered zk-SNARK setup, encrypted identity data, and session-bound nonce verification, which create significant barriers to exploitation. Privileges required are "None," as voters use the system through a public frontend without special access. User interaction is also "None" because all cryptographic processes, including proof generation and submission, are handled automatically. The scope is "Unchanged" since an exploit would not propagate beyond the voting instance. Both confidentiality and integrity impacts are rated "High" due to the critical nature of vote secrecy and correctness. Availability impact is "None" because the system performs verification off-chain and does not rely on continuous blockchain interaction, minimizing disruption from gas cost fluctuations or DoS attempts.

Helios offers strong audit capacity but lacks coercion resistance. Its vulnerabilities include a ballot-secrecy flaw enabling replay attacks that let an adversary cause a detectable election outcome change and learn how the victim voted [15], and a Cross-Site Request Forgery (CSRF) issue where unauthorized commands could be sent from a voter's browser [16]. Helios depends on a registered voter list and authentication (often via email tokens or passwords), exposing it to malicious election admins who could add fake voters or intercept voter credentials. It is also susceptible to coercion, as voters receive a ballot tracking hash as a receipt, public on the bulletin board, allowing them to prove their vote by revealing their ballot and tracker.

Civitas's approach to let people cast fake votes, along with receipt-free interfaces, helps counteract vote selling and coercion attempts. It implies multiple election authorities (trustees), which can act as a weak link in security. One issue is the need for voters to manage secret credentials and collaborate with the trustees, increasing user complexity and leading to mistakes. Its weaknesses are mainly practical: heavy infrastructure, dependence on unusual channels, and no large-scale deployment [17].

Semaphore achieves strong privacy through `zk-SNARK` proofs with each user having a public commitment stored as a leaf in a Merkle tree and a private key that contains a nullifier to prevent double voting. Key problems that render it an inappropriate choice are the gas-intensive protocol for large trees and the fact that the set of eligible voters (the group) is not a secret, adversaries knowing who can vote, but not what they voted.

Our proposed system improves upon existing solutions by combining blockchain-backed audit with `zk-SNARK`-based anonymity. It addresses key vulnerabilities such as coercion and double voting through cryptographic nullifiers and ensures vote integrity via immutable on-chain commitments. Unlike Civitas, it avoids complex credential management by using MetaMask login and automated proof generation, reducing user complexity. Compared to Helios, it eliminates the need for voter-side auditing, and unlike Semaphore, it avoids high gas costs and public group exposure through off-chain proof verification and a hashed eligibility list. However, this layered cryptographic design introduces added complexity and potential risks, including frontend manipulation, MITM attacks, and implementation flaws. Proofs are verified off-chain for efficiency, and only public outputs (nullifier, commitment) are submitted on-chain, balancing performance with a minimal trust assumption.

A key innovation is the separation between eligibility verification (`zk-SNARK-1`) and vote integrity (`zk-SNARK-2`), enabling modular, reusable proofs without revealing sensitive data again. This design reduces circuit complexity, improves transparency, and defends against forgery even if the client is compromised. To our knowledge, this layered `zk-SNARK` structure is unique among blockchain voting systems.

We measured the computational cost of proof generation and verification on a modest system (Intel Core i5-3320M @ 2.60GHz, 8GB RAM). `zk-SNARK-1` proof generation (eligibility circuit) averaged 1.83 s, while `zk-SNARK-2` proof generation (vote circuit) averaged 1.69 s. Verification times were 1.48 s and 1.34 s for `zk-SNARK-1` and `zk-SNARK-2` respectively. These results reflect efficient proof operations even on older consumer hardware, demonstrating the practicality of our architecture for real-world deployment.

An important advantage of our platform is its low operational cost. Compared to national election expenditures, blockchain-based architectures represent only a small fraction of the overall budget. Based on internal assessment and Romanian government spending data, the blockchain-related costs of our system remain under 10% of current spending levels, with contract deployment at 0.001 SepETH and vote verification at 0.0001 SepETH.

5 Conclusions and Future Work

We proposed a flexible, low-cost, transparent and secure alternative to physical elections, achieved through a modular zero-knowledge system that isolates responsibilities and minimizes trust assumptions, with the resulting system demonstrating advantages over existing solutions in areas such as privacy, verifiability, and efficiency. By strengthening security and increasing scalability, as well as significantly leveraging the associated costs for the election process overall, our proposed system encourages a digital trace of the voting process, ensuring its traceability and transparency, while preserving anonymity altogether.

Acknowledgments. This work has been partially supported by (1) project virtuaLedger [18], (2) project RoNaQCI, part of EuroQCI, DIGITAL-2021-QCI-01-DEPLOY-NATIONAL, 101091562, and (3) project "Romanian Hub for Artificial Intelligence - HRIA", Smart Growth, Digitization and Financial Instruments Program, 2021–2027, MySMIS no. 351416

References

1. Goldwasser, S., Micali, S., Rackoff, C.: The knowledge complexity of interactive proof systems. SIAM J. Comput. **18**(1), 186–208 (1989). https://doi.org/10.1137/0218012
2. Blum, M., Feldman, P., Micali, S.: Non-interactive zero-knowledge and its applications. In: Proceedings of the 20th Annual ACM Symposium on Theory of Computing (STOC), pp. 103–112. ACM (1988). https://doi.org/10.1145/62212.62222
3. Ben-Sasson, E., Bentov, I., Horesh, Y., Riabzev, M.: Scalable, transparent, and post-quantum secure computational integrity. Cryptology ePrint Archive, Report 2018/046 (2018). https://eprint.iacr.org/2018/046
4. Groth, J.: On the size of pairing-based non-interactive arguments. In: Fischlin, M., Coron, J.-S. (eds.) EUROCRYPT 2016. LNCS, vol. 9666, pp. 305–326. Springer, Heidelberg (2016). https://doi.org/10.1007/978-3-662-49896-5_11
5. Bellés-Muñoz, M., Isabel, M., Muñoz-Tapia, J.L., Rubio, A., Baylina, J.: Circom: a circuit description language for building zero-knowledge applications. IEEE Trans. Dependable Secure Comput. **20**(6), 4733–4751 (2023). https://doi.org/10.1109/TDSC.2022.3232813
6. HELIOS https://vote.heliosvoting.org/ Accessed 6 May 2025
7. Clarkson, M.R., Chong, S., Myers, A.C.: Civitas: toward a secure voting system. In: 2008 IEEE Symposium on Security and Privacy (SP), pp. 354–368. IEEE (2008). https://doi.org/10.1109/SP.2008.32
8. Juels, A., Catalano, D., Jakobsson, M.: Coercion-resistant electronic elections. In: Proceedings of the 2005 ACM Workshop on Privacy in the Electronic Society (WPES), pp. 61–70. ACM (2005). https://doi.org/10.1145/1102199.1102213
9. Semaphore community: semaphore: zero-knowledge signaling on Ethereum (Whitepaper v1) (2025). https://semaphore.pse.dev/whitepaper-v1.pdf Accessed 22 Apr 2025
10. Nikolaenko, V., Ragsdale, S., Bonneau, J., Boneh, D.: Powers-of-tau to the people: decentralizing setup ceremonies. In: Gennaro, R., Wichs, D. (eds.) Public-Key Cryptography – PKC 2024. LNCS, vol. 14442, pp. 105–134. Springer, Cham (2024). https://doi.org/10.1007/978-3-031-56179-5_5
11. Solidity contributors: solidity documentation (2025). https://docs.soliditylang.org/ Accessed 22 Apr 2025
12. Wood, G.: Ethereum: a secure decentralised generalised transaction ledger. Ethereum Project Yellow Paper **151**, 1–32 (2014). https://ethereum.github.io/yellowpaper/paper.pdf
13. Lee, W.-M.: Using the MetaMask chrome extension. In: *Beginning Ethereum Smart Contracts Programming: With Examples in Python, Solidity, and JavaScript*, pp. 93–126. Apress, Berkeley, CA (2019). https://doi.org/10.1007/978-1-4842-5086-0_5
14. Forum of incident response and security teams (FIRST): common vulnerability scoring system v3.1: user guide (2025). https://www.first.org/cvss/v3-1/user-guide Accessed 27 Apr 2025
15. Cortier, V., Smyth, B.: Attacking and fixing Helios: an analysis of ballot secrecy. Cryptology ePrint Archive, Report 2010/625 (2010). https://eprint.iacr.org/2010/625.pdf

16. Gawel, D., Kosarzecki, M., Vora, P.L., Wu, H., Zagorski, F.: Apollo – end-to-end verifiable internet voting with recovery from vote manipulation. Cryptology ePrint Archive, Report 2016/1037 (2016). https://eprint.iacr.org/2016/1037.pdf
17. Neumann, S., Volkamer, M.: Civitas and the real world: problems and solutions from a practical point of view. In: 2012 Seventh International Conference on Availability, Reliability and Security (ARES), pp. 180–185. IEEE (2012). https://doi.org/10.1109/ARES.2012.75
18. The VirtuaLedger project, https://virtualedger.com Accessed 13 Dec 2024

Study and Comparison of Lattice Sieving Algorithms

Miguel Ángel González de la Torre, Luis Hernández Encinas(✉), and Diego Rojas Rodríguez

Institute of Physical and Information Technologies (ITEFI), Spanish National Research Council (CSIC), C/Serrano 144, 28006 Madrid, Spain
{ma.gonzalez,luis.h.encinas,diego.rojas}@csic.es

Abstract. The Shortest Vector Problem, denoted as SVP, is one of the most important problems related to the security in the post-quantum cryptography, and it is defined over lattices. In this work, we present an state-of-the-art analysis of the different algorithms to solve this problem evaluating their efficiency in High Performance Computing (HPC) systems. To do this, we have modified the Shortest Vector Problem oracle used in the implementation of the Block Korkine-Zolotarev algorithm, known as BKZ, included in the General Sieve Kernel (gsk) software library to test a suite of different algorithms. The main purpose of this research is to check whether current hardness estimations apply to highly parallelized implementations. In this sense, we have tested the different algorithms in a High Performance Computing cluster and compared their performance on highly parallelized environments.

Keywords: Lattice-based Post-quantum Cryptography · Public-Key Cryptography · Shortest Vector Problem · Sieving attacks

1 Introduction

In the context of Post-Quantum Cryptography (PQC) lattice-based cryptosystems are currently considered to be the most promising. Prior to the beginning of the standardisation process by the National Institute of Standards and Technology (NIST), lattice-based cryptography had already been researched for its value in Fully Homomorphic cryptosystems. After the standardization of PQC, there have been several improvements on the attacks and new approaches to tackle the Shortest Vector Problem (SVP), to which many lattice problems can be reduced. In 2024, the first three PQC standards were published, two of which are lattice-based: a Key Encapsulation Mechanism (KEM), ML-KEM [18], and a Digital Signature (DS), ML-DSA [18]. Both standards base their security in the Learning With Errors (LWE) problem, which is considered hard to solve in high dimensions. The LWE problem can be reduced to the SVP, meaning that solving the latter implies also solving the LWE problem.

E. Corchado et al. (Eds.): CISIS 2025, CCIS 2807, pp. 56–67, 2026.
https://doi.org/10.1007/978-3-032-19770-2_6

In this work, our main goal is to study and analyse the different variants of sieving algorithms, used inside the Block Korkine-Zolotarev (BKZ) algorithm [19]. BKZ is the best known attack to lattice cryptography and requires an SVP oracle. It should stay clear that solving the LWE problem with cryptographic parameters is considered to be unfeasible, which does not detract interest in studying other solvable instances of the problem. We will run different sieving oracles in High Performance Computing (HPC) systems where parallelization may allow us to increase performance. We think that optimization of computational resources may lead to a significant reduction on the security of the LWE-based algorithms and thus it should be of upmost importance to verify that resourceful eavesdroppers can not breach the security of what is promised to be the future of information security.

The rest of this communication contains the following sections: Sect. 2 presents the background concepts needed for understanding the rest of this work. Section 3 contains the most relevant strategies to solve the SVP. In Sect. 4 we show the results obtained from the cluster DRAGO, property of the Spanish National Research Council (CSIC), and Finisterrae III, property of Galicia Supercomputing Center (CESGA), we also present the conclusions from our analysis. Finally, in Sect. 5 some possible lines of work to continue the research are included.

2 Finding Short Vectors

A lattice $\mathcal{L}$ is defined as a discrete additive subgroup of $\mathbb{R}^n$. In other words, given a set of linearly independent vectors $B = \{b_0, \ldots, b_{n-1}\} \in \mathbb{R}^n$, we have

Definition 1 (Lattice). *A lattice with basis B is the set of integer linear combinations of the vectors of B: $\mathcal{L}(B) = \{Bx \colon x \in \mathbb{Z}^m\}$.*

The set B is called basis of the lattice and the same lattice can be generated by different basis. In any case, to simplify the notation, the reference to a basis in the definition of a lattice will be omitted unless it is required. A lattice $\mathcal{L} \subset \mathbb{R}^n$ inherits the Euclidean norm, hence there is also a notion of distance between lattice points. The minimum distance of a lattice $\mathcal{L}$ is the length of the shortest nonzero lattice vector: $\lambda_1(\mathcal{L}) := \min_{v \in \mathcal{L} \setminus \{0\}} \|v\|$.

Definition 2 (SVP). *Given an arbitrary basis B of some lattice $\mathcal{L}(B)$, find a shortest nonzero lattice vector, i.e., find $v \in \mathcal{L}(B)$ for which $\|v\| = \lambda_1(\mathcal{L}(B))$.*

For a vector $\mathbf{s} \in \mathbb{Z}_q^n$, called secret, the LWE distribution $A_{\mathbf{s},\chi}$ over $\mathbb{Z}_q^n \times \mathbb{Z}_q$ is defined by choosing $\mathbf{a} \in \mathbb{Z}_q^n$ uniformly at random and $e \leftarrow \chi$ and then sampling $(\mathbf{a}, b = \langle \mathbf{s}, \mathbf{a} \rangle + e \pmod q)$. Once the LWE distribution is defined, the LWE problem can be stated as the following.

Definition 3 (LWE). *Given m independent samples $(\mathbf{a}_i, b_i) \in \mathbb{Z}_q^n \times \mathbb{Z}_q$ from the LWE distribution $A_{\mathbf{s},\chi}$, find $\mathbf{s}$.*

The SVP problem is considered to be in NP under randomized reductions, hence any known algorithms that solve it require exponential time. There are two kinds of algorithms that solve the SVP [17]: approximation algorithms and exact algorithms. The approximation algorithms output a vector that is close, in norm, to the shortest vector; while the exact algorithms probably output the shortest vector. The best examples of probabilistic SVP solvers are the algorithms by Lenstra, Lenstra, and Lovasc (LLL) [15] and the Block Korkin-Zolotarev (BKZ) [19], both are what are called lattice reduction algorithms, i.e. algorithms that replace the basis of the lattice with another one that has better shape and properties. For exact SVP algorithms there are two main families: sieving and enumeration.

As its name suggests, enumeration algorithms aim to calculate all possible solutions to a problem to deterministically find the best solution. More efficient approaches look for sufficiently good solutions and add techniques like pruning the solution tree, reducing significantly the runtime. The estimated cost of enumeration algorithms varies from $2^{O(n^2)}$ to $2^{O(n \log n)}$ [11], depending on the preprocessing of the lattice basis.

Sieving algorithms consists of performing operations inside a list of vectors of the lattice with the intention of obtaining the shortest possible vector. These algorithms require a computational cost of $2^{O(n)}$ and have a space complexity of $2^{O(n)}$. Due to these restrictions, sieving algorithms are hard to implement; however, right now, they are considered as the fastest and optimal choice for solving the exact SVP problem. Currently, sieving algorithms are considered the preferable choice, based on the experimental results from [1,2].

Approximated algorithms like BKZ run an exact SVP solver in a lower dimension (lower than the input dimension), while both sieving and enumeration benefit in their performance of preprocessing the basis of the lattice with algorithms like LLL.

3 Sieving Strategies

The first sieving strategy ever proposed is the AKS sieve, in [1]. However, the implementation that we study corresponds to the NV sieve [17]. It was not until the Gaussian sieve was introduced in [16] that sieving was considered competitive.

Before presenting the different versions of sieving, we will introduce the general concepts related to this type of algorithm. Given vectors $u, v \in \mathcal{L}$, they form a reduced pair if

$$\min\{\|u + v\|, \|u - v\|\} \geq \max\{\|u\|, \|v\|\}.$$

If a pair of vectors is not reduced it is said to be reducible. If two vectors $u, v \in \mathcal{L}$ verify the condition $\theta(u, \pm v) < \pi/3$, then they are a reducible pair, where $\theta(u, v)$ denotes the angle between u and v.

There are three main aspects that change between different implementations: i) how the list of vectors is formed, ii) how large this list has to be to contain reducible pairs in it, and iii) when does the algorithm find a short (sufficiently short) vector. The sampling of the vectors depends on the implementation, but it is advised to output vectors of the lattice following a known and convenient distribution, some examples are [8,12].

The kissing number [10] in dimension d is defined as the maximum number of non-overlapping unit spheres that can touch a single unit sphere. This number is difficult to predict for high dimensions, but there are known bounds to its value. For sieving algorithms, the kissing number is relevant since it also denotes the maximum number of vectors that is possible to consider in dimension d that form angles of $\pi/3$. Consequently, if any list contains a number of vectors higher than the kissing number of the dimension considered, then the existence of a reducible pair is guaranteed. In [10] a lower bound of the kissing number is presented $|\mathcal{L}| \in 2^{0.2075n+o(n)}$.

In AKS a lattice $\mathcal{L}$ of rank n and radius R are taken as input, and generate a list of vectors $S \subset \mathcal{L} \cap B_n(R)$, where $B_n(R)$ denotes the closed ball of radius R and centred in the origin. The value R serves as bound to the length of vectors in the list and is iteratively reduced by a factor δ. In the AKS sieve [1] the following process is followed: Let S be a set, initially empty. Sample vectors from $B_n(R)$ and add them to S to get:

$$S = \{(y_i, v_i) \in B_n(R) \times \mathcal{L} : ||v_i - y_i|| \leq \xi\}, \text{ for a value } \xi \text{ close to } 0.$$

After an iteration, R is reduced to δR, where $\delta \in [2/3, 1)$, and keep the vectors with norm inferior to δR or $(y_i', v_i') = (y_j - v_i, v_j - v_i)$ where $||v_i'|| \leq \delta R$.

3.1 NV Sieve

The NV sieve algorithm [17] is a refined version of the deterministic AKS sieve. The designers of the NV sieve introduce an heuristic approach and reduce significantly the number of parameters involved. Besides the set S of vectors considered in the sieve, two new sets are defined, the set of centres C, initialized as an empty list and the set C_n defined as:

$$C_n(\gamma, R) = \{x \in \mathbb{R}^n : \gamma R \leq ||x|| \leq R\}.$$

The heuristic considered in [17] assumes that vectors in $S \cap C_n(\gamma, R)$ after a sieve iteration are uniformly distributed in $C_n(\gamma, R)$. The main defining characteristic of NV sieving are the initial sampling, the length of the γ factor, and the heuristic distribution of the vectors after each reduction cycle.

While analysing the points in $S \subseteq B_n(R)$, the NV sieve will add vectors with norm greater than γR to the set of centres C. On the contrary, vectors of norm lower than γR are directly introduced in the list of the next iteration. The set C contains points in B_n that could reduce the length of other vectors in S (see Algorithm 1). The amount of iterations is greatly decreased with respect to AKS and, for appropriate values of γ, it returns a set of short enough vectors to solve the γ-SVP. To guarantee the quality of the output one should verify that $|B_n(R) \cap \mathcal{L}| \geq |S|^2$, which gives a good approximation of λ_1. When all the vectors remain in the list of centres during the reduction, the algorithm finishes.

Algorithm 1 NV sieve

```
Require: S ⊆ B_n(R) ∩ 𝓛, 2/3 ≤ γ ≤ 1
Ensure: L ⊆ B_n(γR) ∩ 𝓛
  Initialize C = {}, L = {}
  for all v ∈ S do
      if ||v|| ≤ γR then
          S' = S' ∪ {v}
      else
          if ∃c ∈ C  st ||v − c|| ≤ γR then
              L = L ∪ {v − c}
          else
              C = C ∪ {v}
          end if
      end if
  end for
```

3.2 Gaussian Sieve

The Gaussian approach is similar to the NV sieve, although instead of implementing S as a list it changes it to a queue (following first-in-first-out order). This offers a better space complexity and a better time performance on average. For the operation of removing the first element of a queue and to add an element on the back, the functions *pop* and *push* are defined, respectively.

Most implementations of the Gaussian sieve consider a number of collisions c as stop condition. A collision occurs when the reduction of two different pairs of vectors generate the same outcome. Once a set number of collisions is reached, the algorithm stops and outputs the list of reduced vectors L.

For its part, K increments with each collision of vectors linearly dependent with the ones in the list. Moreover, only vectors in the same $\pi/3$ quadrant are evaluated for subtraction to give vectors with smaller norm. When vectors out of this quadrant are obtained they are rotated to fit in this space. Finally, the function Gaussian.Reduction used in Algorithm 2 is given as the Algorithm 3.

3.3 Triple Sieve

The triple sieve is a particular implementation of a more general construction called tuple-sieve [4]. These algorithms introduce a new strategy in the process of searching for short vectors given a sufficiently long list. While the previously presented sieving algorithms (NV or Gaussian) consider pairs of vectors and analyse whether they are reducible, the k-tuple-sieve type algorithms consider sets of three, four, ... k vectors and look for shortest vectors among the combinations of additions and subtractions of these vectors.

Triple sieve considers operations involving multiple vectors instead of just pairs (see Algorithm 4). This approach asymptotically decreases the size of the list but increases the time of the reduction. The most efficient approach considers triplets of vectors and modifies the Gaussian sieve to operate with them.

Algorithm 2 Gaussian sieve

```
Require: Sample : ∅ → B_n(R)
  Initialize L = {0}, S = {}, K = 0
  while K < c do
      if S ≠ {} then
          v = S.pop()
      else
          v ← Sample
      end if
      v = Gaussian.Reduction(v, L, S)
      if v ≠ 0 then
          L = L ∪ {v}
      else
          K++
      end if
  end while
```

Algorithm 3 Gaussian.Reduction

```
Require: p ← Sample, L ⊂ 𝓛, S ∈ B_n(R)
  while ∃v ∈ L : ||v|| ≤ ||p|| ∧ ||p − v|| ≤ ||p|| do
      p = p − v
  end while
  while ∃v ∈ L : ||v|| > ||p|| ∧ ||v − p|| ≤ ||v|| do
      L = L\{v}
  end while
  return p
```

Algorithm 4 Triple sieve

```
Require: L ⊂ 𝓛, S ∈ B_n(R)
  Initialize L = {0}, S = {}
  while condition do
      if S ≠ {} then
          p = S.pop()
      else
          p = Sample
      end if
      p = Triple.Reduce(p, L, S)
      if p ≠ 0 then
          L = L ∪ p
      end if
  end while
```

Algorithm 5 Triple.Reduce

```
Require: p ← Sample, L ⊂ ℒ, S ∈ B_n(R)
  for all v ∈ L do
    if ||p|| ≥ ||v|| ∧ ||p − v|| ≤ ||p|| then
      p = p − v
    end if
  end for
  for all v ∈ L do
    if ||v|| ≥ ||p|| ∧ ||v − p|| ≤ ||v|| then
      v = v − p
      S = S.push(v)
    end if
  end for
  for all v_1, v_2 ∈ L do
    if ||p − v_1 − v_2|| ≤ ||p|| then
      p = p − v_1 − v_2
    end if
  end for
  for all v_1, v_2 ∈ L do
    if ||v_1 − p − v_2|| ≤ ||v_1|| then
      v_1 = v_1 − p − v_2
      S = S.push(v_1)
    end if
  end for
  return p
```

If $p = 0$ at any time in this process, then one could return 0 to avoid non necessary operations. The function Triple.Reduce called in Algorithm 4 is given as the Algorithm 5.

3.4 List Decoding Sieve

Locally sensitive sieving algorithms use additional preprocessing to increase the probability of success in the search of reducible pairs once the list of vectors is sampled. In particular, there has been two different preprocessing methods introduced to sieving, Locality Sensitive Hashing (LSH) and Locality Sensitive Filtering (LSF).

Locality Sensitive Hashing. The idea behind LSH is to store the vectors of the sieve in hash tables. The hash functions used for the application of LSH to sieving consists of functions that map any n-dimensional vector $v \in R^n$ to a low-dimensional sketch and the probability of two vectors having the same sketch grows as the vectors are closer.

This technique was first introduced to solve the Approximate Nearest Neighbors Search (NNS) problem in [9]. Later, in [14], it is used LSH applied to sieving to solve the NNS problem in high dimensions and in the cases that the list L has a certain structure. Moreover, in [6], it is applied the LSH sieve to improve the performance of the Gaussian sieve and NV sieve.

Although in [13] is introduced LSH to both, NV and Gaussian sieves, this kind of improvement has been mostly applied to the latter.

Locality-Sensitive Filtering. Let consider the same definition as LSH, but instead of using the hash output to sort the vectors of L, it is consider now a set of filters $\{f_i\}$. These filters are defined as binary mappings: each vector either pass or does not pass the filter. In [5] the authors define a set of filters defined by spherical caps. The spherical cap of angle θ and centre $u \in B_n(R)$ is defined as $C^{n-1}(\theta, u) := \{v \in B_n(R) : angle(u, v) \leq \theta\}$. Basically, applying a filter $f := \{(x_i, \theta_i)\}$ to the list of vectors L considered in the sieving means calculating $L \cap C^{n-1}(\theta_i, x_i)$.

4 Results in HPC

The implementations of the sieving algorithms we have used in our tests are from the General Sieve Kernel (g6k) [2], which is an open source library under the GNU General Public Licence (GPL) https://github.com/fplll/g6k. This kernel gives the basic operations needed to run algorithms like LLL, BKZ, and multiple implementations of sieving. All the sieving defined in the previous section are included in g6k. The triple sieve implemented in g6k is denoted as hk3, while the LSF sieve in g6k is the version introduced in [5], called bgj1. The internal structure relies on buckets where sieving is performed independently and after the reduction is completed a central database is updated with a list of shortest vectors on each bucket. There is not yet any successful implementation that heavily relies in parallelization, often key in HPC. As shown in [3], in dimension 120 and above, the parallelism using the MPI (Message Passing Interface) standard is around the 80%. In the implementation we have considered, the parallelization is used to open different buckets with a single thread on each bucket.

Apart from the software layer, the hardware becomes quite relevant to run sieving on high dimensions. The exponential cost on space may increase to the order of terabytes of RAM, worsened by the multithreading. Moreover, the greatest limiting factor has become the bandwidth of the RAM because, as we have mentioned, when a bucket is successfully reduced the thread has to add the shortest vectors to the database. This saturates the memory and reduces the performance overall.

Our initial result is the comparison between sieving and the other SotA exact SVP algorithm, enumeration Figs. 1 and 2. The main purpose of this comparison is to verify what has been previously established by the scientific community: that sieving algorithms are currently the best strategy for solving shortest vector type problems.

It can be clearly appreciated in Fig. 1 that for dimension higher than 70 sieving is faster than enumeration. Although for lower dimensions it is still considered enumeration, these cases are far from being cryptographically relevant, hence for any estimation of the cost of solving the SVP problem the logical consideration are sieving algorithms.

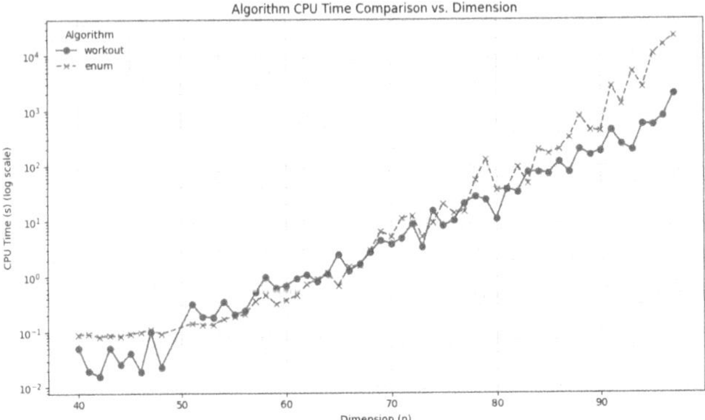

Fig. 1. Experimental results after running different sieves in DRAGO.

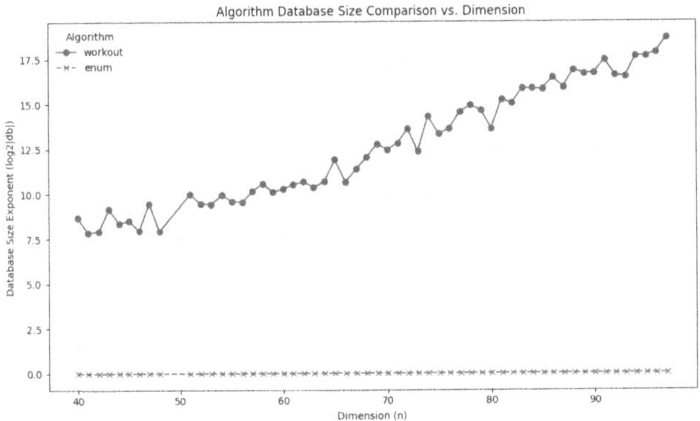

Fig. 2. Comparison between sieving and enumeration.

In Figs. 3 and 4 it can be appreciated a comparison between the different presented sieving strategies, in particular of the implementation in g6k of these sieves.

In Fig. 3 it can be appreciated that the hk3 and bgj1 sieves present a better performance in time than the NV or Gaussian sieves, as the dimension considered for the sieve is higher. In terms of memory, as can be seen in Fig. 4, the Gaussian sieve shows the better performance. However, the difference in the case of the memory usage does not seem to grow with the dimension, while in the time it does.

We would also want to highlight performance advantage that HPC systems offer to solve these problems. The instances of the SVP problem we solved were proposed in the SVP lattice challenge https://www.latticechallenge.org/svp-challenge/ are considered.

In Table 1 we compare previously published results from the aforementioned implementation using MPI (∗ data not included in [3]), the native g6k and our

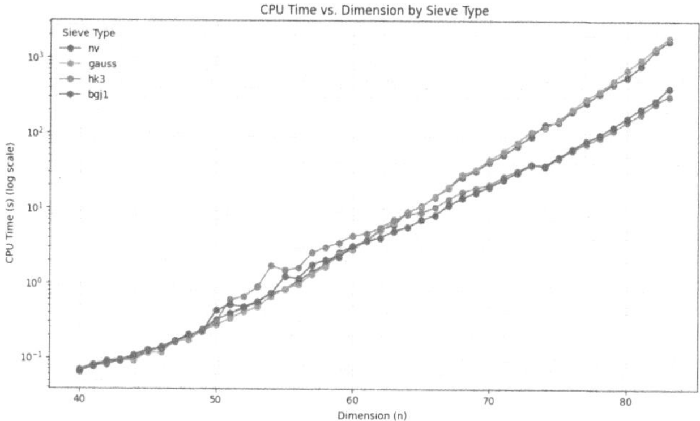

Fig. 3. Time used by different sieving in increasing dimension.

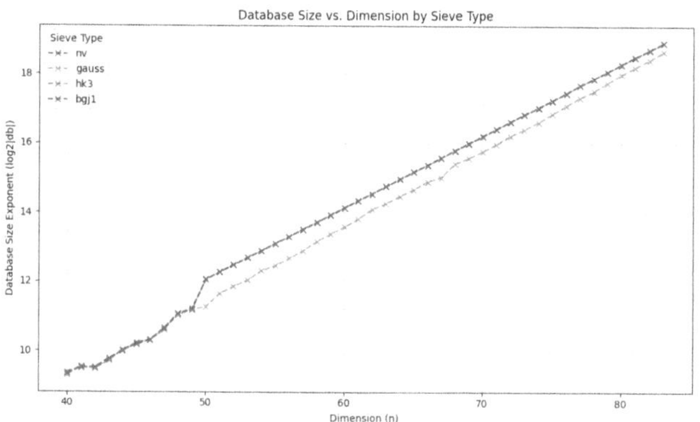

Fig. 4. Memory used by different sieving in increasing dimension.

Table 1. Comparison of performance between sieving executions

g6k version	SVP dim	Max Sieve dim	CPU time (hours)	Memory usage (GB)
Native g6k [2]	128	102	95	7,6
GPU g6k [7]	158	129	382	90
MPI g6k [3]	128	*	102	320
Finisterrae	130	103	24	7,54

results obtained in the Finisterrae III supercomputer. Additionally we offer a SotA reference based in CUDA architecture for GPUs. It can be appreciated how we reached the same dimension as the original publication in almost one forth of the time and obtain significantly better results than the MPI implementation.

5 Future Work

To the knowledge of the authors, there are no implementations that successfully implements multithreading inside of the buckets. Adapting the code of G6K to use MPI to do this would be of great interest to see whether there are diminishing returns on the amount of threads working on a task or if sieving algorithms with enough resources could pose a threat to LWE algorithms.

There also are implementations that use CUDA architecture for speeding up the reduction of pairs of vectors obtaining state-of-the-art results. Having access to clusters with high performance GPU would be interesting to complement our results. Furthermore, we have not reached cryptographic dimensions, but only some toy-sized examples. The estimations suggest that our cluster would not have enough space to fit this cases, but exploring bigger systems could allow us to consider solving problems on larger dimensions.

Acknowledgments. This work was supported by Project Quantum-based Resistant Architectures and Techniques Integration QKD+PQC (QURSA) funded by MCIN/AEI/10.13039/501100011033, co-funded by EU "NextGenerationEU"/PRTR under Grant TED2021-130369B-C33.

Compliance with ethical standards

Disclosure of Interests. M.A.G.T. wishes to thank CSIC for its support under the EFiDiP project, and D.R.R. thanks to the JAE ICU 2024 ITEFI-CSIC for its support.

References

1. Ajtai, M., Kumar, R., Sivakumar, D.: A sieve algorithm for the shortest lattice vector problem. In: Proceedings of the 33rd Annual ACM Symposium on Theory of Computing, pp. 601–610. ACM (2001). https://doi.org/10.1145/380752.380857
2. Albrecht, M.R., Ducas, L., Herold, G., Kirshanova, E., Postlethwaite, E.W., Stevens, M.: The general sieve kernel and new records in lattice reduction. Cryptology ePrint Archive, Paper 2019/089 (2019). https://eprint.iacr.org/2019/089
3. Albrecht, M.R., Rowell, J.: Scaling lattice sieves across multiple machines. Cryptology ePrint Archive, Paper 2024/747 (2024). https://eprint.iacr.org/2024/747
4. Bai, S., Laarhoven, T., Stehlé, D.: Tuple lattice sieving. LMS J. Comput. Math. **19**(A), 146–162 (2016). https://doi.org/10.1112/S1461157016000292
5. Becker, A., Ducas, L., Gama, N., Laarhoven, T.: New directions in nearest neighbor searching with applications to lattice sieving. Cryptology ePrint Archive, Paper 2015/1128 (2015). https://eprint.iacr.org/2015/1128
6. Becker, A., Laarhoven, T.: Efficient (ideal) lattice sieving using cross-polytope LSH. In: Pointcheval, D., Nitaj, A., Rachidi, T. (eds.) Progress in Cryptology – AFRICACRYPT 2016. AFRICACRYPT 2016. LNCS, vol. 9646, pp. 3–23. Springer, Cham (2016). https://doi.org/10.1007/978-3-319-31517-1_1
7. Ducas, L., Stevens, M., van Woerden, W.: Advanced lattice sieving on gpus, with tensor cores. In: Canteaut, A., Standaert, F.X. (eds.) Advances in Cryptology – EUROCRYPT 2021. EUROCRYPT 2021. LNCS, vol. 12697, pp. 249–279. Springer, Cham (2021). https://doi.org/10.1007/978-3-030-77886-6_9

8. Gentry, C., Peikert, C., Vaikuntanathan, V.: Trapdoors for hard lattices and new cryptographic constructions. In: Proceedings of the Fortieth Annual ACM Symposium on Theory of Computing, pp. 197–206. STOC '08, Association for Computing Machinery, New York, NY, USA (2008). https://doi.org/10.1145/1374376.1374407
9. Indyk, P., Motwani, R.: Approximate nearest neighbors: towards removing the curse of dimensionality. In: Proceedings of the Thirtieth Annual ACM Symposium on Theory of Computing, pp. 604–613. STOC'98, Association for Computing Machinery, New York, NY, USA (1998). https://doi.org/10.1145/276698.276876
10. Jenssen, M., Joos, F., Perkins, W.: On kissing numbers and spherical codes inhigh dimensions. Adv. Math. **335**, 307–321 (2018). https://doi.org/10.1016/j.aim.2018.07.001
11. Kannan, R.: Improved algorithms for integer programming and related lattice problems. In: Proceedings of the Fifteenth Annual ACM Symposium on Theory of Computing, pp. 193–206. STOC'83, Association for Computing Machinery, New York, NY, USA (1983). https://doi.org/10.1145/800061.808749
12. Klein, P.: Finding the closest lattice vector when it's unusually close. In: Proceedings of the Eleventh Annual ACM-SIAM Symposium on Discrete Algorithms, pp. 937–941. SODA '00, Society for Industrial and Applied Mathematics, USA (2000)
13. Laarhoven, T.: Sieving for shortest vectors in lattices using angular locality–sensitive hashing. In: Gennaro, R., Robshaw, M. (eds.) Advances in Cryptology – CRYPTO 2015, LNCS, pp. 3–22. Springer, Berlin, Heidelberg (2015). https://doi.org/10.1007/978-3-662-47989-6_1
14. Laarhoven, T., de Weger, B.: Faster sieving for shortest lattice vectors using spherical locality–sensitive hashing. In: Lauter, K., Rodríguez-Henríquez, F. (eds.) Progress in Cryptology – LATINCRYPT 2015, LNCS, pp. 101–118. Springer, Cham (2015). https://doi.org/10.1007/978-3-319-22174-8_6
15. Lenstra, A., Lenstra, H., Lovasz, L.: Factoring polynomials with rational coefficients. Mathematische Annalen **261**(4), 515–534 (1982). https://doi.org/10.1007/BF01457454
16. Micciancio, D., Voulgaris, P.: Faster exponential time algorithms for theshortest vector problem. In: Proceedings of the 2010 Annual ACM–SIAM Symposium on Discrete Algorithms (SODA), pp. 1468–1480. https://doi.org/10.1137/1.9781611973075.119
17. Nguyen, P.Q., Vidick, T.: Sieve algorithms for the shortest vector problem are practical. J. Math. Cryptol. **2**(2), 181–207 (2008). https://doi.org/10.1515/JMC.2008.009
18. NIST: Module-Lattice-Based Digital Signature Standard. National Institute of Standard and Technology, NIST FIPS 204 (2023). https://doi.org/10.6028/NIST.FIPS.204.ipd
19. Schnorr, C., Euchner, M.: Lattice basis reduction: Improved practical algorithms and solving subset sum problems. Math. Programm. **66**(1–3), 181–199 (1994). https://doi.org/10.1007/BF01581144

Hierarchical Threshold Structure-Preserving Signatures

Ahmet Ramazan Ağırtaş[2], Emircan Çelik[1(✉)], and Oğuz Yayla[1]

[1] Institute of Applied Mathematics, Middle East Technical University, 06800 Çankaya, Ankara, Turkey
emircancelik92@gmail.com, oguz@metu.edu.tr
[2] AYWARE, 06420 Çankaya, Ankara, Turkey

Abstract. We introduce a digital signature scheme that combines hierarchical secret sharing with threshold structure-preserving signatures. In this scheme, a global signing key is recursively partitioned into layers, each protected by its own threshold, so that distinct subsets of the same participant group can exercise role-specific signing authority. Hierarchical Threshold Structure-Preserving Signatures (HTSPS) distribute two SPS signing keys across a hierarchy by interpreting each user's share as a polynomial-derivative value, so higher-level users hold evaluations and lower-level users hold higher-order derivatives; any group of participants that meets the vector of thresholds can interpolate to obtain a valid SPS signature, while sub-threshold sets see only random group elements. Our design composes Tassa's Birkhoff-based share system with the threshold SPS yielding signature protocol with a multi-layered access structure.

Keywords: Structure-Preserving Signatures · Threshold Signatures · Hierarchical Secret Sharing Keys

1 Introduction

Structure Preserving Signatures (SPS) are pairing-based syntax compatible digital signature schemes, meaning, all the messages, public keys and the signatures are elements of the same bilinear group elements that the protocol operates in. This enables the reduction of verification to pairing-product equations within the group itself. This useful property allows a modular and efficient protocol design; therefore, there has been a lot of research interest on the topic.

Abe et al. initiated this research by introducing the concept in [1]. In 2016, Ghadafi [12] proposed a shorter SPS than existing SPS schemes. Its randomizable key variant with an adaptable algorithm is given in [8]. In 2019, Crites and Lysyanskaya [7] built mercurial signatures that are malleable with respect to the message, signature, and key space, focusing on Structure-Preserving Signatures on Equivalence Classes (SPS-EQ) [13]. Later, Connolly et al. [5] proposed a new variant of it based on the assumption of the common reference string (CRS)

E. Corchado et al. (Eds.): CISIS 2025, CCIS 2807, pp. 68–77, 2026.
https://doi.org/10.1007/978-3-032-19770-2_7

model. Inspired by SPS-EQ, Backes et al. [4] presented signatures featuring a flexible public key primitive that converts the key into an alternative representative of the same equivalence class. Recently, Mir et al. [14] proposed the first aggregatable SPS-EQ, introducing aggregate signatures with randomizable public key and tags, along with aggregate mercurial signatures. In 2024, Abe et al. [15] presented interactive threshold mercurial signatures that expand the scope of threshold SPS to encompass EQ.

Many SPS in the literature are inadequate for thresholding due to nonlinear processes or even require significant overhead. However, in 2023, Crites et al. [6] proposed a Threshold Structure-Preserving Signature (TSPS) by defining an indexed Diffie-Hellman message space. They address thresholdization by employing indexing that converts each scalar message m into an index id to generate partial signatures in a compatible format for aggregation in a threshold scheme. Furthermore, the authors in [3] extend this scheme with a randomizable key.

Assume a scenario that there is an organization with multi-level of hieararchy. Assume that this organization is willing to create a policy where the authority of a top-level official must be shared within the ranks while keeping the hierarchical structure. TSPS uses standard secret sharing scheme by Shamir secret-sharing scheme that split a secret into pieces within a group of participants so that only authorized subgroups of participants can recover it. The classical example is Shamir's (t, n)-threshold construction [16], which treats every participant symmetrically by encoding the secret as the free coefficient of a random polynomial over a finite field. Tassa extends this scheme into Hierarchical Secret Sharing (HSS) [17] with enabling different levels of ranking among participants in terms of hierarchy. Tassa's scheme realizes such access structure within our scenario while remaining perfect and ideal by handing higher-level users evaluations of the polynomial and lower-level users evaluations of its higher-order derivatives; reconstruction is then a Birkhoff interpolation that succeeds if and only if combiner group meets every threshold.

In this paper, we will introduce Hierarchical Threshold Structure Preserving Signatures, an extension of Threshold Structure Preserving Signature scheme with combined ability to establish a complex access structure of a secret with different levels of authoritarianity. Outline of this paper as follows; in Sect. 2 we will lay out the necessary definitions and notation, in Sect. 3 we will introduce our contribution and give the correctness proof and finally Sect. 4 mention conclusion remarks and future work.

2 Preliminaries

In this section we will introduce necessary concepts and notation to build our protocol. For the syntax, λ is the security parameter 1^λ is its unary representation, and $negl(\lambda)$ denotes negligible function. While $x \xleftarrow{\$} X$ is utilized to indicate that x is uniformly sampled from the set X, the cardinality of a set X is represented by the symbol $|X|$. Also, $\mathsf{A}(\mathsf{x}) \rightarrow \mathsf{y}$ denotes that y is an output of the algorithm A on input x, whereas $\mathsf{x} \leftarrow \mathsf{y}$ denotes the straightforward assignment process. We denote linear pairing function e as described in [11].

Definition 1 (Bilinear Map). *A bilinear map is a function $e : \mathbb{G}_1 \times \mathbb{G}_2 \rightarrow \mathbb{G}_T$ in groups $\mathbb{G}_1, \mathbb{G}_2, \mathbb{G}_T$ with prime order p, that satisfies:*

- *Bilinearity: $e(g_1^m, g_2^n) = e(g_1, g_2)^{m.n}$ for all $m, n \in \mathbb{Z}_p$, $g_1 \in \mathbb{G}_1$, $g_2 \in \mathbb{G}_2$.*
- *Non-degeneracy: $e(g_1, g_2) \neq 1$ for all $g_1 \in \mathbb{G}_1$, $g_2 \in \mathbb{G}_2$.*

The pairing map is termed symmetric (Type I) if $\mathbb{G}_1 = \mathbb{G}_2$. Asymmetric pairing occurs when $\mathbb{G}_1 \neq \mathbb{G}_2$. In this scenario, if an effectively computable isomorphism $\phi : \mathbb{G}_2 \rightarrow \mathbb{G}_1$ exists, it is categorized as Type II; otherwise, it is classified as Type III, which will be also relied on in this study.

Definition 2 (Diffie-Hellman Message Space $\mathcal{M}_{DH}$ [10]). *Given a Type III bilinear map $(\mathbb{G}_1, \mathbb{G}_2, \mathbb{G}_T, p, e, g_1, g_2)$. A pair $(M_1, M_2) \in \mathbb{G}_1 \times \mathbb{G}_2$ belongs to the Diffie-Hellman message space $\mathcal{M}_{DH}$ if there exist $m \in Z_p$ such that $M_1 = g_1^m$ and $M_2 = g_2^m$. Verification of $e(M_1, g_2) = e(g_1, M_2)$ determines such a pair.*

Definition 3 (Indexed Diffie-Hellman Message Space $\mathcal{M}_{iDH}^H$ [6]). *Given a Type III bilinear map $(\mathbb{G}_1, \mathbb{G}_2, \mathbb{G}_T, p, e, g_1, g_2)$, an index set $\mathcal{I}$ ($id \in \mathcal{I}$), and a random oracle $\mathsf{H} : \mathcal{I} \rightarrow \mathbb{G}_1$. $\mathcal{M}_{iDH}^H$ is an Indexed Diffie-Hellman message space if the followings hold:*

(1) $\mathcal{M}_{iDH}^H \subset \{(id, \tilde{M}) | id \in \mathcal{I}, m \in \mathbb{Z}_p, \tilde{M} = (\mathsf{H}(id)^m, g_2^m) \in \mathbb{G}_1 \times \mathbb{G}_2\}$
(2) For all $(id, \tilde{M}) \in \mathcal{M}_{iDH}^H$, $(id', \tilde{M}') \in \mathcal{M}_{iDH}^H$, $id = id'$, then $\tilde{M} = \tilde{M}'$.

Definition 4 (Shamir's Secret Sharing [16]). *Shamir's Secret Sharing is a (t, n)-threshold scheme that enables the division of a secret s into n shares such that any subset of shares containing at least t shares can reconstruct the secret s, while any subset containing fewer than t shares cannot obtain any knowledge regarding s. Consider a finite field of prime order p, denoted as $\mathbb{F}_p$. Shamir's secret sharing comprises share generation and secret reconstruction as outlined below:*

- *$\mathsf{ShareGen}(s, p, n, t)$: The dealer selects a random polynomial $f(x) \in \mathbb{F}_p[x]$ of degree $t - 1 < p$, where coefficients selected from $\mathbb{F}_p$, such that $f(0) = s \in \mathbb{F}_p$. The dealer secretly transmits a secret share $s_i = f(id_i) \mod p$ to participant P_i, where $id_i \in \mathbb{F}_p$ denotes the identifier of P_i, and the vector of shares is $\boldsymbol{s} = (s_1, \ldots, s_n)$.*
- *$\mathsf{Reconst}(\{s_i\}_{i \in \mathcal{T}})$: The secret is reconstructed by Lagrange interpolation on the polynomial $f(x)$ as $s = f(0) = \sum_{i \in \mathcal{T}} s_i \cdot \lambda_i$, where $\mathcal{T} \subseteq [1, n]$ of size at least t, and the Lagrange coefficient $\lambda_i = \prod_{j \in \mathcal{T}, j \neq i} \frac{id_j}{id_j - id_i} \in \mathbb{F}_p$.*

Definition 5 (Feldman's Verifiable Secret Sharing). *Feldman's Verifiable secret sharing [9] enables users to check the consistency of the received shares. Assume we have n players setup, let q, p be primes such that $q \mid p - 1$ and $g \in \mathbb{F}_p$*

be an element of order q. In this scheme, dealer first selects a secret s to be shared and forms the random polynomial $f(x) \in \mathbb{F}_q[x]$ of degree $t-1 < q$

$$f(x) = a_0 + a_1 x + a_2 x^2 + \ldots + a_{t-1} x^{t-1}$$

where $a_i \in \mathbb{F}_q$, for $i = 0, 1, \ldots t-1$ with $a_0 = s$. Then, dealer computes the commitments to these coefficients by $C_i = g^{a_i}$ for $i = 0, 1, \ldots, t-1$ and publishes commitment values. The dealer sends the share $f(j)$ to each participant P_j for $j = 1, \ldots, n$. After receiving its share, participant P_j confirms the consistency of the received share by checking the equality

$$g^{f(j)} = \prod_{k=0}^{t-1} C_k^{j^k}.$$

If the equality holds, participants can decide that no malicious interruption has occurred and the share is consistent with the secret s. The reconstruction of the secret follows the same steps as in Shamir's secret sharing scheme.

Definition 6 (Hierarchical Threshold Secret Sharing Scheme [17]). *Let $\mathcal{G}$ be set of n players such that it is established by m distinct subsets, i.e., $\mathcal{G} = \bigcup_{i=1}^{m} \mathcal{G}_i$, $\mathcal{G}_i \cap \mathcal{G}_j = \emptyset$ for $i \neq j$. Let $0 < k_1 < \ldots < k_m$ be the degree of the access structure levels. Then, the access structure Γ of the hierarchical threshold secret sharing scheme is*

$$\Gamma = \left\{ \mathcal{V} \subset \mathcal{G} : \left| \mathcal{V} \cap \left(\bigcup_{j=1}^{i} \mathcal{G}_j \right) \right| \geq k_i \mid \forall i \in \{1, 2, \ldots, m\} \right\}$$

Dealer then samples a random polynomial of degree k_m, i.e.,

$$f(x) = \sum_{i=0}^{k_m} a_i x^i$$

with $a_0 = s$. The dealer assigns a unique element $x_u \in \mathbb{F}_q$ to each user $u \in \mathcal{G}_i$ and sends the share $(x_u, f^{(k_{i-1}+1)})$, i.e., $f^{(k_{i-1}+1)}$ is the $k_{i-1}+1$-th derivative of polynomial f, where $k_0 = -1$. In the secret construction, Birkhoff coefficients are used, which are described next.

Definition 7 (Birkhoff Coefficient). *[2, Def. 2.9] Let $X = \{x_1, x_2, \ldots, x_k\}$ be a set of given points in $\mathbb{R}$ where $x_1 < x_2 < \ldots < x_k$, $E = (e_{i,j})$ for $i = 0, \ldots, k$ and $j = 0, \ldots, \ell$ be a matrix with binary entries, $I(E) = \{(i,j) : e_{i,j} = 1\}$, $d = |I(E)|$, and $C = \{c_{i,j} : (i,j) \in I(E)\}$ be a set of d real values (we assume hereafter that the right-most column in E is nonzero). Let matrix A be defined by $A(E, X, \phi_j) = (\theta_{ij})_{d \times d}$ where $\theta_{ij} = g_{j-1}^{(\alpha_i(2))}(x_{\alpha_i(1)})$ for $i, j = 1, \ldots, d$. Birkhoff coefficient β_i is the evaluation of the polynomial f_i at 0,*

$$f_i(x) = \sum_{j=0}^{d-1} (-1)^{(i+j)} \frac{\det(A_i(E, X, \phi_j))}{\det(A(E, X, \phi))} g_j(x),$$

i.e. $\beta_i = f_i(0)$.

3 Hierarchical Threshold Structure Preserving Signatures

In this section, we present a Hierarchical Threshold Structure Preserving Signature Scheme (HTSPS) which extends Threshold Structure Preserving Scheme with a hierarchical access structure. With HTSPS, one can define a complex access-structure setting in which delegation is possible both vertically and horizontally. Since the higher level of shares includes more information in terms of derivatives, one can delegate this responsibility downward without the need of changing the global verification key. This also naturally enables different valid subgroups in the same access levels.

Imagine the following scenario; a multinational investment bank for legal and risk-management reasons has a policy that mandates the following policy; any issuance of an account above some level requires three-tier approval ladder which includes; approval from Tier 0 group; which could be an example of executive board, approval from Tier 1 group; which could be an example of risk and compliance board and approval from Tier 2 group; which could be an example of regional directors. HTSPS can provide a setting where this policy's requirements are met while any sub-coalition that does not meet the threshold for each tier learns nothing about the global signature.

HTSPS consists of six PPT algorithms, that is, Setup, KeyGen, $\mathsf{PSignGen}$, $\mathsf{PSignVer}$, $\mathsf{Reconst}$, Verify. Since HTSPS uses HTSS for secret key generation, this scheme also relies on a designated dealer to setup the access structure and generation of secret keys and verification keys.

- $\mathsf{Setup}(1^\lambda) \to pp$: The setup algorithm accepts the security parameter 1^λ as an input and outputs the public parameters pp for the HTSPS. The public parameters consist of groups $\mathbb{G}_1, \mathbb{G}_2, \mathbb{G}_T$ of orders p used in bilinear map e, along with the generators g_1 and g_2 for $\mathbb{G}_1$, and $\mathbb{G}_2$, respectively. Additionally, there is a hash function $\mathsf{H} : \{0,1\}^m \to \mathbb{G}_1$ that maps to $\mathbb{G}_1$ allowing messages to be securely represented as elements within the group. Algorithm outputs the pubic parameters $pp := ((\mathbb{G}_1, \mathbb{G}_2, \mathbb{G}_T, p, e, g_1, g_2), \mathsf{H})$
- $\mathsf{KeyGen}(pp) \to (sk_1, sk_2, vk_1, vk_2)$: The key generation algorithm is used by the selected dealer to generate key pair that will be used to generate signature. The selected dealer proceeds to establish the access structure as in Definition 6, i.e., let $\mathcal{G}$ be set of n participant such that it is established by m distinct subsets, i.e., $\mathcal{G} = \bigcup_{i=1}^m \mathcal{G}_i$, $\mathcal{G}_i \cap \mathcal{G}_j = \emptyset$ for $i \neq j$. Let $0 < k_1 < \ldots < k_m$ be the degree of the access structure levels and let n_i be the number of participant in the corresponding level $\mathcal{G}_i$. The access structure Γ is defined by

$$\Gamma = \left\{ \mathcal{V} \subset \mathcal{G} : \left| \mathcal{V} \cap \left(\bigcup_{j=1}^{i} \mathcal{G}_j \right) \right| \geq k_i \mid \forall i \in \{1, 2, \ldots, m\} \right\}$$

  With the access structure established the dealer proceeds as follows;
  - Dealer chooses two random secret values $sk_1, sk_2 \in \mathbb{F}_p$ and calculates global verification keys $vk_1 := g_2^{sk_1}$ and $vk_2 := g_2^{sk_2}$.

- Dealer samples two polynomials $f_1(x), f_2(x) \in \mathbb{F}_p[x]$ of degree k_m

$$f_1(x) = \sum_{i=0}^{k_m} a_i x^i, \quad f_2(x) = \sum_{i=0}^{k_m} b_i x^i$$

  with coefficients $a_i, b_i, i = 1, \ldots, k_m$ selected uniformly random such that $sk_1 = a_0$ and $sk_2 = b_0$ as seen in Fig. 1.
- Dealer calculates and publishes the commitments $B_i = g_2^{a_i}$ and $C_i = g_2^{b_i}$ for $i = 1, \ldots, k_m$.
- Dealer assigns a unique identifier $x_{ij} \in \mathbb{F}_p$ to each participant $P_{ij} \in \mathcal{G}_i$ for $i \in [1, m]$ and $j \in [1, n_i]$.
- Dealer calculates the shares by taking the $k_{i-1} + 1$-th derivatives of both f_1 and f_2, and sends the values to the corresponding participant, i.e., each participant receives $(x_{ij}, \psi_{ij,1}, \psi_{ij,2})$, where $\psi_{ij,1} = f_1^{(k_{i-1}+1)}(x_{ij})$ and $\psi_{ij,2} = f_2^{(k_{i-1}+1)}(x_{ij})$.
- After receiving its share $(x_{ij}, \psi_{ij,1}, \psi_{ij,2})$ each participant first does consistency check by

$$g_2^{\psi_{ij,1}} = \prod_{j=k_i}^{k_m} B_j^{\binom{j}{j-k_i} x_{ij}^{j-k_i}} \quad \text{and } g_2^{\psi_{ij,1}} = \prod_{j=k_i}^{k_m} C_j^{\binom{j}{j-k_i} x_{ij}^{j-k_i}},$$

  where k_i is the access level the participant P_{ij} belongs to. If both of the equations hold then participant accepts the shares, otherwise it rejects it.
- Each participant $P_{ij} \in \mathcal{G}_i$ for $i \in [1, m]$ and $j \in [1, n_i]$ calculates their partial verification key pair $(vk_{ij,1}, vk_{ij,2})$ by $vk_{ij,1} = g_2^{\psi_{ij,1}}$, $vk_{ij,2} = g_2^{\psi_{ij,2}}$.

– $\mathsf{PSignGen}(pp, x_{ij}, \psi_{ij,1}, \psi_{ij,2}, (id, M_1, M_2)) \to (\sigma_{ij}, \perp)$: Partial signature generation algorithm is used by each participant that generates partial signature values.
  – Participants perform the indexed message consistency check on indexed message $(id, M_1, M_2) \in \mathcal{M}_{iDH}^H$, by

$$h \neq 1_{\mathbb{G}_1} \wedge M_1 \neq 1_{\mathbb{G}_1} \wedge e(h, M_1) = e(M_1, g_2)$$

  with message $\mu \in \mathbb{F}_p$, where $h = \mathsf{H}(id)$, $M_1 = h^\mu$ and $M_2 = g_2^\mu$.
  – Each participant calculates its own partial signature σ_{ij}

$$\sigma_{ij} = (h, s_{ij}) = (h, h^{\psi_{ij,1}} \cdot M_1^{\psi_{ij,2}}),$$

  and sends it to the combiner who can be any one of the signers.

$\mathsf{PSignVer}(pp, (id_i, M_{1,i}, M_{2,i}), \sigma_{ij} \to \{0, 1\})$: Partial signature verification algorithm is used to verify each participant P_{ij}'s partial signature. The combiner accepts the received partial signature if all the following conditions hold:

(i) $h_i \neq 1_{\mathbb{G}_1}$,

(ii) $M_{1,i} \neq 1_{\mathbb{G}_1}$,
(iii) $e(h_i, M_{2,i}) = e(M_{1,i}, g_2)$,
(iv) $e(h_i, vk_{ij,1})e(M_{1,i}, vk_{ij,2}) = e(s_{ij}, g_2)$.
Rejects it otherwise.

- $\mathsf{Reconst}(pp, \mathcal{S}, \{\sigma_{1j}, \sigma_{2j}, \ldots, \sigma_{mj}\}_{j=1}^{n_i}) \rightarrow \sigma$: The reconstitution algorithm is used by the valid access structure to reconstruct the signature. The combiner first establishes the subgroup of signers $\mathcal{S} := \{\mathcal{S}_1, \ldots, \mathcal{S}_m\}$ where $\mathcal{S}_i$ is the subgroup of signers in respective access levels such that $|\mathcal{S}_i| > k_i$. The combiner then calculates the signature by

$$\sigma = (h, s) = (h, \prod_{i=1}^{m} \prod_{j \in \mathcal{S}_i} s_{ij}^{\beta_{ij}})$$

where β_{ij} is the corresponding Birkhoff coefficient defined in Definition 7.

- $\mathtt{Verify}(pp, \sigma, (id, M_1, M_2), vk_1, vk_2) \rightarrow \{0, 1\}$: Verify algorithm is used to verify the signature of the message with index id. The verifier checks the validity of the following equations,
(i) $h \neq 1_{\mathbb{G}_1}$,
(ii) $M_1 \neq 1_{\mathbb{G}_1}$,
(iii) $e(h, M_2) = e(M_1, g_2)$,
(iv) $e(h, vk_1)e(M_1, vk_2) = e(s, g_2)$.
If the equations hold the verifier accepts the signature, otherwise rejects it.

3.1 Security and Correctness

We can show that the scheme is correct by

$$\begin{aligned}
s &= \prod_{i=1}^{m} \prod_{j \in \mathcal{S}_i} s_{ij}^{\beta_{ij}} \\
&= \prod_{i=1}^{m} \prod_{j \in \mathcal{S}_i} (h^{\psi_{ij,1}} \cdot M_1^{\psi_{ij,2}})^{\beta_{i,j}} \\
&= \prod_{i=1}^{m} \prod_{j \in \mathcal{S}_i} (h^{\psi_{ij,1}\beta_{i,j}} \cdot M_1^{\psi_{ij,2}\beta_{i,j}}) \\
&= h^{\sum_{i=1}^{m} \sum_{j \in S_i} \beta_{i,j}\psi_{ij,1}} \cdot M_1^{\sum_{i=1}^{m} \sum_{j \in S_i} \beta_{i,j}\psi_{ij,2}} \\
&= h^{sk_1} \cdot M_1^{sk_2}
\end{aligned}$$

It's easy to see that the above equation is true for signature σ. Our scheme inherits two main security features directly from its building blocks. First, the confidentiality follows the perfect security of Tassa's hierarchical threshold secret-sharing scheme: any group of participants that falls short of the vector $k_0, k_1, \ldots, k_m$ learns nothing about the signing key information-theoretically, because their shares leave at least one coefficient of the secret polynomial

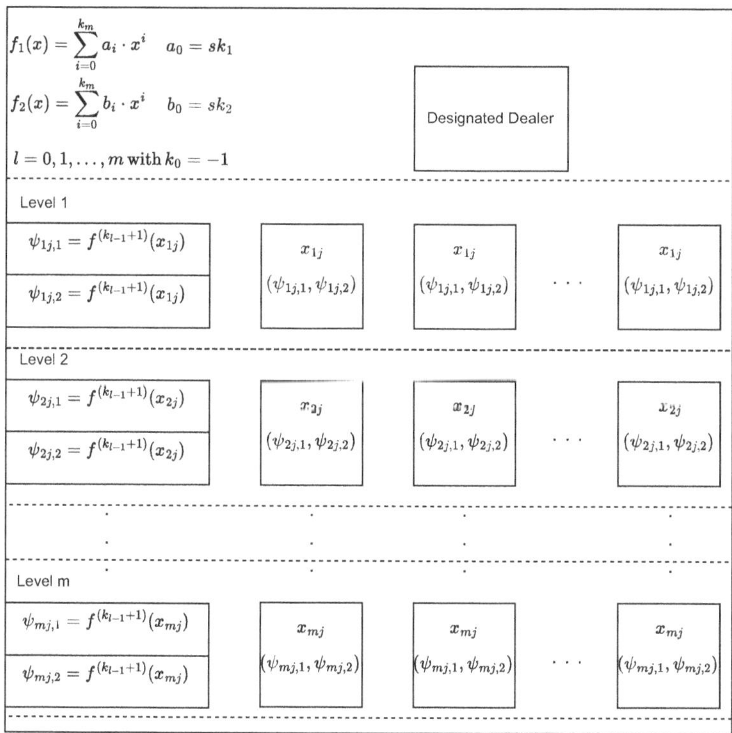

Fig. 1. Share Distribution in HTSPS.

undetermined. Second, existential unforgeability under chosen-message attack reduces the security of TSPS: forging an aggregate signature without meeting the threshold would either reveal a non-authorized linear combination of the secret shares, contradicting the security of HTSS, or produce a valid SPS signature without the underlying signing key, breaking the hardness assumption of TSPS. Since our aggregation of signatures is a linear combination in the exponent, it introduces no additional algebraic structure. Thus, the overall security of our protocol is the intersection of HTSS confidentiality and TSPS unforgeability.

4 Conclusion and Future Work

In this paper, we presented Hierarchical Threshold Structure Preserving Signatures where we extend threshold structure preserving signatures with a hierarchical secret sharing scheme, allowing the protocol to enable a layered access structure. The proof of correctness is also provided, and we argue that the security of our scheme does not separate from the security of the building blocks HTSS and TSPS. Our protocol still relies on the designated dealer to set up the access structure and the combiner to aggregate the signature. As future work, we plan to provide a proof-of-concept implementation and an efficiency analysis of our proposed scheme. Also, the formal proof of the scheme will be done as future work.

References

1. Abe, M., Fuchsbauer, G., Groth, J., Haralambiev, K., Ohkubo, M.: Structure-preserving signatures and commitments to group elements. In: Rabin, T. (ed.) Advances in Cryptology – CRYPTO 2010. CRYPTO 2010. LNCS, vol. 6223, pp. 209–236. Springer, Berlin, Heidelberg (2010). https://doi.org/10.1007/978-3-642-14623-7_12
2. Ağırtaş, A.R., Yayla, O.: Compartment-based and hierarchical threshold delegated verifiable accountable subgroup multi-signatures. In: Dąbrowski, A., Pieprzyk, J., Pomykała, J. (eds.) Number-Theoretic Methods in Cryptology, LNCS, vol. 14966, pp. 283–313. Springer, Cham (2025). https://doi.org/10.1007/978-3-031-82380-0_10
3. Ağırtaş, A.R., Çelik, E., Kocaman, S., Sulak, F., Yayla, O.: Threshold structure-preserving signatures with randomizable key. In: Proceedings of the 22nd International Conference on Security and Cryptography, pp. 632–637 (2025)
4. Backes, M., Hanzlik, L., Kluczniak, K., Schneider, J.: Signatures with flexible public key: introducing equivalence classes for public keys. In: Peyrin, T., Galbraith, S. (eds.) Advances in Cryptology – ASIACRYPT 2018. ASIACRYPT 2018. LNCS, vol 11273, pp. 405–434. Springer, Cham (2018). https://doi.org/10.1007/978-3-030-03329-3_14
5. Connolly, A., Lafourcade, P., Perez Kempner, O.: Improved constructions of anonymous credentials from structure-preserving signatures on equivalence classes. In: Hanaoka, G., Shikata, J., Watanabe, Y. (eds.) Public-Key Cryptography – PKC 2022. PKC 2022. LNCS, vol 13177, pp. 409–438. Springer, Cham (2022). https://doi.org/10.1007/978-3-030-97121-2_15
6. Crites, E., Kohlweiss, M., Preneel, B., Sedaghat, M., Slamanig, D.: Threshold structure-preserving signatures. In: Guo, J., Steinfeld, R. (eds.) Advances in Cryptology – ASIACRYPT 2023. ASIACRYPT 2023. LNCS, vol. 14439, pp. 348–382. Springer, Singapore (2023). https://doi.org/10.1007/978-981-99-8724-5_11
7. Crites, E.C., Lysyanskaya, A.: Delegatable anonymous credentials from mercurial signatures. In: Matsui, M. (ed.) Topics in Cryptology – CT-RSA 2019. CT-RSA 2019. LNCS, vol. 11405, pp. 535–555. Springer, Cham (2019). https://doi.org/10.1007/978-3-030-12612-4_27
8. Derler, D., Slamanig, D.: Key-homomorphic signatures: definitions and applications to multiparty signatures and non-interactive zero-knowledge. Des. Codes Cryptogr. **87**, 1373–1413 (2019)
9. Feldman, P.: A practical scheme for non-interactive verifiable secret sharing. In: 28th Annual Symposium on Foundations of Computer Science (sfcs 1987), pp. 427–438. IEEE (1987)
10. Fuchsbauer, G.: Automorphic signatures in bilinear groups and an application to round-optimal blind signatures. Cryptology ePrint Archive (2009)
11. Galbraith, S.D., Paterson, K.G., Smart, N.P.: Pairings for cryptographers. Discret. Appl. Math. **156**(16), 3113–3121 (2008). https://doi.org/10.1016/j.dam.2007.12.010, https://www.sciencedirect.com/science/article/pii/S0166218X08000449, applications of Algebra to Cryptography

12. Ghadafi, E.: Short structure-preserving signatures. In: Sako, K. (ed.) Topics in Cryptology - CT-RSA 2016. CT-RSA 2016. LNCS, vol. 9610, pp. 305–321. Springer, Cham (2016). https://doi.org/10.1007/978-3-319-29485-8_18
13. Hanser, C., Slamanig, D.: Structure-preserving signatures on equivalence classes and their application to anonymous credentials. In: Sarkar, P., Iwata, T. (eds.) Advances in Cryptology – ASIACRYPT 2014. ASIACRYPT 2014. LNCS, vol. 8873, pp. 491–511. Springer, Berlin, Heidelberg (2014). https://doi.org/10.1007/978-3-662-45611-8_26
14. Mir, O., Bauer, B., Griffy, S., Lysyanskaya, A., Slamanig, D.: Aggregate signatures with versatile randomization and issuer-hiding multi-authority anonymous credentials. In: Proceedings of the 2023 ACM SIGSAC Conference on Computer and Communications Security, pp. 30–44 (2023)
15. Nanri, M., Kempner, O.P., Tibouchi, M., Abe, M.: Interactive threshold mercurial signatures and applications. Cryptology ePrint Archive (2024)
16. Shamir, A.: How to share a secret. Commun. ACM **22**(11), 612–613 (1979)
17. Tassa, T.: Hierarchical threshold secret sharing. In: Naor, M. (ed.) Theory of Cryptography. TCC 2004. LNCS, vol. 2951, pp. 473–490. Springer, Berlin, Heidelberg (2004). https://doi.org/10.1007/978-3-540-24638-1_26

A Comparative Study of LWE and LIP-Based Post-Quantum Signature Schemes

Édgar Pérez-Ramos(✉), Cristian Nina-Orellana, Candelaria Hernández-Goya, and Pino Caballero-Gil

University of La Laguna, Tenerife, Spain
{eperezra,alu0101470856,mchgoya,pcaballe}@ull.edu.es

Abstract. With the rise of quantum computing as a real threat to classical cryptography, the National Institute of Standards and Technology launched a standardization process in 2016 to identify secure algorithms for public-key encryption and digital signatures. In 2022, CRYSTALS-Kyber was selected for encryption and CRYSTALS-Dilithium, FALCON, and SPHINCS+ were chosen for digital signatures, the first three based on lattice problems and the last on cryptographic hash functions. A second call was initiated in 2024 to encourage diversity in digital signature schemes. In 2025, the candidate HAWK emerged in the second round, offering a lattice-based approach built on the Lattice Isomorphism Problem, a less explored but promising alternative. This work presents and explores two of the central mathematical problems underlying these digital signature algorithms: the Learning With Errors problem, which supports many of the currently standardized algorithms, and the Lattice Isomorphism Problem, which forms the basis of the HAWK proposal. After introducing and analyzing these problems, we implement and compare the signature schemes to evaluate their practical performance. This comparative study aims to analyze the trade-off between digital signatures constructed from cryptographic hash functions and those derived from hard problems on lattices, particularly highlighting the differences between schemes built on the Learning With Errors problem and those based on the Lattice Isomorphism Problem.

Keywords: Post-quantum cryptography · HAWK · Lattices · LIP

1 Introduction

The revolution of quantum computing promises a technological paradigm shift by solving in mere seconds mathematical problems that would take today's most powerful computers thousands of years to compute. However, alongside these advancements comes a significant threat to cybersecurity. Quantum computing could be used to break current encryption systems that safeguard our banking data, communications, and government secrets. This has triggered a race against

E. Corchado et al. (Eds.): CISIS 2025, CCIS 2807, pp. 78–87, 2026.
https://doi.org/10.1007/978-3-032-19770-2_8

time to develop new cryptographic methods that are resilient in the quantum era, ensuring a secure digital future.

Since 2016, the National Institute of Standards and Technology (NIST) has devoted considerable effort to the selection of algorithms that are resistant to quantum attacks. At the end of 2017, 69 algorithms passed the first phase of the call. Later, in 2019, after a thorough screening process, only 26 remained. During that round, rigorous tests were conducted to evaluate the security and performance of each algorithm, as well as to identify potential vulnerabilities.

In the third round of evaluation in 2018, 15 final candidates were selected to continue in the standardization process. During this round, additional testing was conducted, and collaboration with the cryptographic community helped identify potential vulnerabilities and improve the robustness of shortlisted algorithms.

Finally, in July 2022, NIST announced the algorithms selected as final standards. The standardization proposal includes:

- Encryption: CRYSTALS-Kyber [1].
- Digital signatures: CRYSTALS-Dilithium [2], FALCON [3], and SPHINCS+ [4].

Furthermore, in mid-2022, the BIKE, Classic McEliece, HQC (Hamming Quasi-Cyclic), and SIKE encryption schemes advanced to the fourth round, to continue their analysis and evaluation [5].

Regarding digital signatures, in 2023, the next step for NIST was to diversify its range of schemes, including some alternatives not based on lattices [6]. This strategy responds to the principle of crypto-agility, as it ensures the ability of security systems to quickly adapt to changes in cryptographic algorithms, promoting the coexistence of different cryptographic families as a preventive measure against future vulnerabilities. In particular, the search for schemes with short signatures and fast verification was prioritized, thus favouring their application in resource-constrained contexts or those with high efficiency requirements. In 2024, NIST published the second round of this process and released the algorithms that had advanced to that phase [7]. These algorithms are shown in Table 1.

Table 1. Digital signature schemes selected for the second round of the NIST Additional Signature Schemes Evaluation Process (2024).

Category	Algorithms
Code-Based	CROSS, LESS
Lattice-Based	HAWK
MPC-in-the-Head	MIRA*, MiRitH*, MQOM, PERK, RYDE, SDitH
Multivariate	MAYO, QR-UOV, SNOVA, UOV
Symmetric-Based	FAEST
Isogeny-Based	SQIsign

This work is organized as follows. Section 2 introduces the computational problems under analysis: the Learning With Errors (LWE) problem and the Lattice Isomorphism Problem (LIP). It also presents a comparative study of current standardized digital signatures based on lattices and hash functions. Section 3 focuses on the implementation and presents the results obtained, comparing lattice-based signatures on one hand and the HAWK signature scheme against current standardized signatures on the other. Finally, the last section is dedicated to future work and concluding remarks.

2 Background

In this section, we will address the main problems associated with the digital signatures discussed in this work: Learning With Errors and the Lattice Isomorphism Problem. We focus on these problems because we are interested in comparing the efficiency of signature schemes based on LWE and LIP. Both problems are foundational in post-quantum cryptography, and analyzing them will allow us to assess their advantages and disadvantages in terms of security and performance in the context of quantum-resistant digital signatures.

The following notation will be used throughout the paper:

- The ring of integers modulo a prime q is denoted as $\mathbb{Z}_q$.
- The set of n-vectors over $\mathbb{Z}_q$ is denoted as $\mathbb{Z}_q^n$.
- The polynomial ring $\mathbb{Z}_q(x)/\phi(x)$ is denoted as $\mathcal{R}_q$, where $\phi(x)$ is the polynomial x^n+1.

2.1 The Learning With Errors Problem

Problem Formalization. Let $n \in \mathbb{N}$ and $q \in \mathbb{Z}$. Consider m vectors $b_1, b_2, \ldots, b_m \in \mathbb{Z}_q^n$. The lattice Λ generated by the base of vectors $B = \{b_1, b_2, \ldots, b_m \in \mathbb{Z}_q^n\}$ is then the set:

$$\Lambda = \mathcal{L}(b_1, \ldots, b_m) = \left\{ \sum_{i=1}^{m} z_i \cdot b_i \ : \ z_i \in \mathbb{Z} \right\} \tag{1}$$

Fix a probability distribution $\mathcal{X}$ over $\mathbb{Z}_q$, which allows selecting an error term in a controlled manner. The selection of an error e according to this distribution is denoted by $e \leftarrow \mathcal{X}$. These parameters define the so-called *LWE distribution* over $\mathbb{Z}_q^n \times \mathbb{Z}_q$.

Definition 1. *LWE Distribution. Let $n, q \in \mathbb{N}$, $s \in \mathbb{Z}_q^n$, and $\mathcal{X}$ be a probability distribution over $\mathbb{Z}_q$. The LWE distribution modulo q associated with s, denoted $\mathcal{A}_{s,\mathcal{Z}}$, is defined by selecting a vector $a \in \mathbb{Z}_q$ uniformly at random and selecting an error $e \leftarrow \mathcal{X}$, yielding:*

$$(a, b) \in \mathbb{Z}_q^n \times \mathbb{Z}_q, \quad \textit{where } b = \langle s, a \rangle + e \mod q \tag{2}$$

Applications of *LWE* to Cryptography. This section is based on [8]. As mentioned earlier, a wide variety of cryptographic constructions have been based on the standard LWE problem. Most of these applications can be made more efficient, and sometimes even practical for real-world use, by adapting them to Ring-LWE (RLWE). This adaptation process is usually straightforward, though in some cases it requires additional technical tools to achieve the most precise and efficient results.

As an example of application, a simple and efficient semantically secure public-key cryptosystem is outlined. The key generation algorithm selects a uniformly random element $A \in \mathcal{R}_q$ as well as two small random elements, $s, e \in \mathcal{R}_q$, from the LWE distribution. Then, s is used as the secret key, and the pair $(A, t = A \cdot s + e)$ forms the public key. To encrypt a message m of n bits, where $m = (b_0, b_1, \ldots, b_{n-1})$ with $b_i \in \{0, 1\}$, the encryption algorithm selects three small random elements $r, e_1, e_2 \in \mathcal{R}_q$ from the error distribution, and the corresponding calculations produce the pair (u, v) as the ciphertext of the message m, where:

$$u = A \cdot r + e_1 \mod q, \qquad v = t \cdot r + e_2 + [\frac{q}{2}] \cdot m \mod q \tag{3}$$

To decrypt, we compute:

$$v - u \cdot s = (r \cdot e - s \cdot e_1 + e_2) + [\frac{q}{2}] \cdot m \mod q \tag{4}$$

For an appropriate choice of parameters, the coefficients of $r \cdot e - s \cdot e_1 + e_2$ have magnitudes smaller than $\frac{q}{4}$, so the bits of m can be recovered by rounding each coefficient of $v - u \cdot s$ to either 0 or $[\frac{q}{2}]$. Once rounded, those that are 0 remain 0, and those that are $[\frac{q}{2}]$ are set to 1, thus recovering the original message.

Correctness of LWE. According to [8], with n being the security parameter, m the number of equations, q the chosen prime modulus, and a noise parameter $\alpha > 0$, Regev recommends choosing q in the range $(n^2, 2n^2)$, $m = n \cdot \log(q)$, and $\alpha = \frac{1}{n \cdot \log^2(n)}$. Notice that if it weren't for the error in LWE samples, $t - A \cdot s$ could be 0 or $[\frac{q}{2}]$ depending on the encrypted bit, and decryption would always be correct. Thus, Regev indicates that a decryption error only occurs if the sum of the error terms across all coefficients exceeds $\frac{q}{4}$. Since we are summing at most n normal error terms, each with a standard deviation of $\alpha \cdot q$, the standard deviation of the sum is at most $\sqrt{m} \cdot \alpha \cdot q << \frac{q}{\log(n)}$; a standard calculation shows that the probability of this normal variable exceeding $\frac{q}{4}$ is negligible.

2.2 The Lattice Isomorphism Problem

The Lattice Isomorphism Problem [9], is a computational problem that arises in the context of lattice theory and cryptography. LIP is the underlying computational problem of the candidate digital signature standard HAWK. It involves determining whether two given lattices are isomorphic, meaning that there exists an orthonormal transformation that maps one lattice onto the other. The difficulty of solving this problem is closely related to the hardness of lattice problems, which are believed to remain hard even in the presence of quantum computers.

The LIP, while less studied than LWE, holds significant cryptographic potential. In work [10], the authors introduce generalizations that enable a worst-case to average-case reduction for the search-LIP within a specific class of isomorphisms. They also present a key encapsulation mechanism and a digital signature scheme built upon the hardness of LIP.

Furthermore, in the more recent work [11] (2025), they prove the feasibility of constructing a fully homomorphic encryption (FHE) scheme based on a variant of the lattice isomorphism problem. Altogether, these developments indicate that although LIP remains relatively underexplored, it is emerging as a promising foundation for post-quantum cryptography, especially driven by the rise of HAWK.

In order to define this problem, we first need to introduce the concept of isomorphism between lattices.

Definition 2. *Let $O \in M_{n\times n}(\mathbb{R})$. We say that O is orthonormal if $O^T O = I_n$.*

Definition 3. *Two lattices $\mathcal{L}_1$ and $\mathcal{L}_2$ are isomorphic if there exists an orthonormal transformation $O \in O_n(\mathbb{R})$ such that $O \cdot \mathcal{L}_1 = \mathcal{L}_2$.*

Next, the search version of lattice isomorphism problem is to determine if two lattices are isomorphic and to find the corresponding orthonormal transformation:

Definition 4 (*decision Lattice Isomorphism Problem (dLIP)*). *Given two lattices $\mathcal{L}_1$ and $\mathcal{L}_2$, determine if there exists a transformation $O \in O_n(\mathbb{R})$ such that:*

$$\mathcal{L}_2 = O \cdot \mathcal{L}_1 \tag{5}$$

Similarly, we define the search version of the problem:

Definition 5 (*search Lattice Isomorphism Problem (sLIP)*). *Given two lattices $\mathcal{L}_1$ and $\mathcal{L}_2$, find, if it exists, the transformation $O \in O_n(\mathbb{R})$ such that:*

$$\mathcal{L}_2 = O \cdot \mathcal{L}_1 \tag{6}$$

Now, considering the bases of the lattices $\mathcal{L}_1$ and $\mathcal{L}_2$, denoted as B_1 and B_2, respectively, we can translate the problem into the following relation between bases: $\mathcal{L}_1$ and $\mathcal{L}_2$ are isomorphic if there exists a unimodular matrix $\mathbf{U} \in GL_n(\mathbb{Z})$ and an orthonormal transformation $O \in O_n(\mathbb{R})$ such that $B_2 = O \cdot B_1 \cdot U$.

Note 1. It is important to recall that two bases B and B' generate the same lattice if there exists a unimodular matrix $\mathbf{U} \in GL_n(\mathbb{Z})$ such that: $B' = B \cdot U$.

The utility of the Lattice Isomorphism Problem in the HAWK scheme is to leverage the difficulty of the modular isomorphism problem in its search version. Specifically, we start with the canonical basis $B = I_n$, which generates $\mathbb{Z}^n$, and consider an obfuscated basis of the form $B \cdot U$. However, unlike in the Shortest Vector Problem (SVP), we also consider an orthonormal transformation (such as a rotation). By constructing a new basis $O \cdot B \cdot U$, where $U \in GL_n(\mathbb{Z})$ is unimodular and $O \in O_n(\mathbb{R})$ is orthonormal, we generate a new lattice that is isomorphic to the original. Since this new basis involves a rotation, it hides the relationship with the canonical basis. This ensures a clear separation between the public information and the secret key.

We have introduced two core hardness assumptions: the Learning With Errors problem and the Lattice Isomorphism Problem, which underlie the ML-DSA and HAWK schemes, respectively. Both problems rely on the assumed difficulty of certain tasks in lattice-based cryptography, yet they differ significantly in structure and algebraic formulation. Additionally, we consider the Short Integer Solution (SIS) problem, which can be seen as a variant of LWE and serves as the foundation for the Falcon signature scheme. Finally, the hash collision resistance problem, which underpins SPHINCS+, has been thoroughly analyzed in existing literature and is well understood. With all these foundational problems established and their differences clarified, we are now prepared to implement the corresponding digital signature schemes and carry out a comparative analysis. The schemes considered are summarized in Table 2.

Table 2. Comparison of Post-Quantum Signature Schemes

Scheme	Underlying Structure	Hardness Assumption
Lattice-based: HAWK		
HAWK-256	Lattice	Lattice Isomorphism Problem
HAWK-512	Lattice	Lattice Isomorphism Problem
HAWK-1024	Lattice	Lattice Isomorphism Problem
Lattice-based: ML-DSA		
ML-DSA-44	Lattice	Module-LWE
ML-DSA-65	Lattice	Module-LWE
ML-DSA-87	Lattice	Module-LWE
Lattice-based: Falcon		
Falcon-512	Lattice	NTRU-SIS
Falcon-1024	Lattice	NTRU-SIS
Hash-based: SPHINCS+		
SPHINCS-128s	Hash-based	Collision Resistance
SPHINCS-128f	Hash-based	Collision Resistance
SPHINCS-256f	Hash-based	Collision Resistance

3 Implementation and Experimental Analysis

For this experiment, we utilized a computer with the following specifications:

- Processor: AMD Ryzen™ 7 4800H, 2.90 GHz (base frequency)
- Processor cores: 8 physical cores/16 threads
- RAM memory: 16.0 GB DDR4

The software stack consists of:

- liboqs: A C library for quantum-resistant cryptographic algorithms from the Open Quantum Safe project.
- The c implementation of HAWK [12], as Open Quantum Safe does not yet contain it in its repositories.

The objective of these experiments is twofold. First, we aim to compare current lattice-based digital signature standards, such as ML-DSA and FALCON, with the candidate standard HAWK. This comparison also provides insight into the extent to which differences arise from the underlying lattice problems-namely Learning With Errors and the Lattice Isomorphism Problem. Second, we seek to evaluate the digital signature schemes SPHINCS+ and ML-DSA against HAWK, providing an additional perspective by assessing the candidate in relation to both lattice-based and hash-based approaches.

For these experiments, we conducted two sets of tests- one comparing LWE-based and LIP-based schemes, and another comparing lattice-based and hash-based digital signatures. In each case, we evaluated the schemes at two different security levels: the lowest and the highest available. Specifically, we used ML-DSA-44, HAWK-256, FALCON-512, and SPHINCS+-128 s for the lower security level, and ML-DSA-87, HAWK-1024, FALCON-1024, and SPHINCS+-256f for the higher level. The results are presented below:

- In the first comparison between LWE and LIP at the lowest security level (see Table 3), the most significant differences arise during the key generation phase. Here, we observe discrepancies of one to two orders of magnitude compared to ML-DSA-44, FALCON, and HAWK, with HAWK being the slowest. In the signing phase, performance differences are minimal across all schemes. Similarly, during verification, the differences remain small, although slightly more noticeable than in the signing phase, yet still far less than in key generation. At the highest security level, the differences follow a similar pattern to those observed in the lowest level. These results are more clearly illustrated in Fig. 1. In this case, due to the large disparities in magnitude, a logarithmic scale is used to enhance visualization.
- In the second comparison (lattice-based vs. hash-based) at the lowest security level (see Table 4), the most significant differences arise during the key generation phase. HAWK-256 shows the highest key generation time, being several orders of magnitude slower than both ML-DSA-44 and SPHINCS-128 s. In the signing and verification phases, performance differences are minimal. At the highest security level, the results follow a similar pattern to those observed

at the lower level. As in first comparison, a logarithmic scale has also been applied in the graphical representation of Fig. 1 due to the significant differences in value magnitudes.

Table 3. Average execution times.

Algorithm	KeyGen (μs)	Sign (μs)	Verify (μs)
Low Security Level			
FALCON-512	6.21×10^3	4.43×10^2	4.27×10^1
HAWK-256	1.54×10^4	4.22×10^2	4.34×10^2
ML-DSA-44	2.03×10^2	4.51×10^2	9.17×10^1
High Security Level			
FALCON-1024	1.68×10^4	5.84×10^2	7.34×10^1
HAWK-1024	6.67×10^4	8.59×10^2	8.69×10^2
ML-DSA-87	3.25×10^2	7.81×10^2	2.27×10^2

Table 4. Average execution times.

Algorithm	KeyGen (μs)	Sign (μs)	Verify (μs)
Low Security Level			
HAWK-256	1.46×10^4	3.94×10^2	4.06×10^2
ML-DSA-44	2.11×10^2	4.82×10^2	9.49×10^1
SPHINCS+-128s	2.22×10^3	4.86×10^2	2.88×10^3
High Security Level			
HAWK-1024	6.53×10^4	5.84×10^2	7.74×10^1
ML-DSA-87	6.66×10^4	8.59×10^2	8.69×10^2
SPHINCS+-256f	3.25×10^2	7.81×10^2	2.27×10^2

Overall, these experiments gave us a clearer view of how different cryptographic foundations behave in practice. They offer valuable insight into the current role of HAWK in the post-quantum landscape and help frame the discussion that follows.

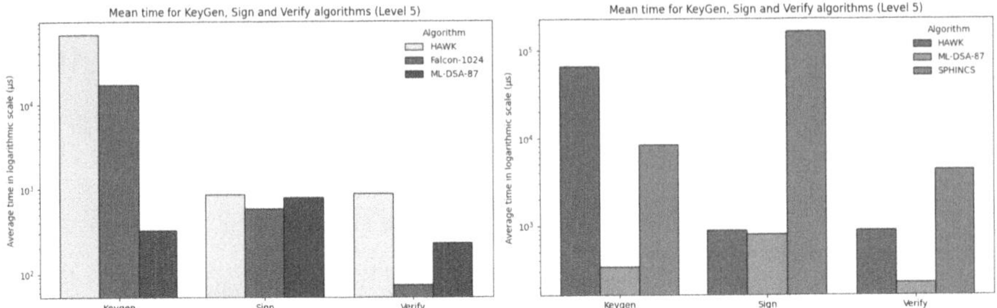

Fig. 1. Mean execution time (in logarithmic scale) for KeyGen, Sign, and Verify operations at the highest security level (Level 5).

4 Conclusions

This work has presented an in-depth study of the current landscape of post-quantum cryptography, with a particular focus on digital signature schemes. We began by analyzing the mathematical foundations of the LWE problem, which underpins many standardized schemes. In parallel, we explored the LIP, a less-explored yet promising alternative forming the basis of the HAWK proposal.

Based on these theoretical foundations, we implemented and evaluated four signature algorithms: ML-DSA, FALCON, SPHINCS+, and HAWK. Two sets of comparative experiments were conducted. The first compared ML-DSA, FALCON, and HAWK, aiming to highlight performance differences between LWE-based and LIP-based lattice schemes. The second compared ML-DSA, SPHINCS+, and HAWK, to position HAWK against currently standardized signature algorithms.

In the first comparison, HAWK demonstrated significantly slower key generation times, while performing similarly to others during the signing phase and showing a slight advantage in verification. However, its high cost in key generation, combined with the absence of notable gains in other phases, limits its practical appeal as an efficient lattice-based signature scheme.

The second comparison further reinforced this observation. Although HAWK outperforms SPHINCS+ in signing and verification phases, it still lags behind ML-DSA, which consistently offers superior overall performance. These results suggest that HAWK, while competitive in some aspects, does not currently present a compelling efficiency advantage over existing standards.

From an algebraic standpoint, HAWK remains an intriguing candidate, as it brings visibility to the LIP, a structurally rich and relatively unexplored problem in lattice-based cryptography. Nonetheless, our findings indicate that HAWK may not be the most efficient choice for digital signatures. Future efforts might be better directed toward diversifying the space of signature schemes, exploring alternatives based on isogenies or multivariate constructions, such as SQIsign, MAYO, or UOV, which show promise both in theory and practice.

Acknowledgments. This work was possible thanks to the projects: 2023DIG28 IACTA, PID2022-138933OB-I00 ATQUE, and SCITALA C064/23 ULL-INCIBE, and to the C065/23 Cybersecurity Chair of the University of La Laguna and INCIBE, funded by Cajacanarias la Caixa Fundations, MCIN/AEI/10.13039/501100011033, and the Recovery, Transformation, and Resilience Plan (Next Generation) financed by the European Union.

References

1. Avanzi, R., et al.: CRYSTALS-Kyber algorithm specifications and supporting documentation. NIST PQC Round **2**(4), 1–43 (2021)
2. Ducas, L., et al.: CRYSTALS-Dilithium: algorithm specifications and supporting documentation (version 3.1). NIST Post-Quantum Cryptogr. Stand. Round **3**, 1–38 (2021)
3. Fouque, P.A., et al.: Falcon: Fast-fourier lattice-based compact signatures over ntru. Submiss. NIST's Post-Quantum Cryptogr. Stand. Process **36**(5), 1–75 (2018)
4. Aumasson, J.P., et al.: SPHINCS+ Specification: Submission to the NIST Post-Quantum Cryptography Standardization Project (Round 3.1). https://sphincs.org/data/sphincs+-r3.1-specification.pdf (2021). disponible en:
5. National Institute of Standards and Technology (NIST): Post-quantum cryptography, round 4 submissions. https://csrc.nist.gov/projects/post-quantum-cryptography/round-4-submissions
6. National Institute of Standards and Technology: Post-quantum cryptography: Additional digital signature schemes. https://csrc.nist.gov/projects/pqc-dig-sig/round-2-additional-signatures
7. Alagic, G., et al.: Status report on the first round of the additional digital signature schemes for the nist post-quantum cryptography standardization process. Tech. Rep. NIST IR 8528, National Institute of Standards and Technology, Gaithersburg, MD (2024). https://doi.org/10.6028/NIST.IR.8528
8. Regev, O.: The learning with errors problem (invited survey). In: Proceedings of the 2010 IEEE 25th Annual Conference on Computational Complexity, pp. 191–204. CCC '10, IEEE Computer Society, USA (2010). https://doi.org/10.1109/CCC.2010.26
9. Haviv, I., Regev, O.: On the lattice isomorphism problem. In: Proceedings of the Twenty-Fifth Annual ACM-SIAM Symposium on Discrete Algorithms, pp. 391–404. SIAM (2014)
10. Ducas, L., van Woerden, W.: On the lattice isomorphism problem, quadratic forms, remarkable lattices, and cryptography. In: Dunkelman, O., Dziembowski, S. (eds.) Advances in Cryptology – EUROCRYPT 2022. EUROCRYPT 2022. LNCS, vol. 13277, pp. 643–673. Springer, Cham (2022). https://doi.org/10.1007/978-3-031-07082-2_23
11. Branco, P., Malavolta, G., Maradni, Z.: Fully-homomorphic encryption from lattice isomorphism. Cryptology ePrint Archive (2025)
12. Pornin, T., Pulles, L.: Hawk implementation. https://github.com/hawk-sign/dev

AI Safety, Privacy and Trustworthy Systems

The Effectiveness of Personal Data Detection in LLM-Based Conversational Agents

Diego Paracuellos(✉), Jose Such, Elena Del Val, and Ana Garcia-Fornes

VRAIN, Universitat Politècnica de València, 46022 Valencia, Spain
diepade@etsinf.upv.es

Abstract. The growing prevalence of LLM-based conversational agents in everyday applications has led to an increasing risk of users disclosing sensitive personal information. Understanding how effectively different tools can identify such disclosures, and therefore protect users, is critical to mitigate privacy risks in human-agent interactions. This paper aims to evaluate the effectiveness of different methods to detect personal information in human-agent conversations. In particular, we compare the potential of several out-of-the-box LLMs as detection agents to more traditional approaches such as Microsoft Presidio. To do so, we use a labeled dataset containing various human interactions with conversational agents. We show that both approaches have strengths and weaknesses, and that none of them on their own seem effective enough to detect personal information in human-agent interactions in uncontrolled, real-world environments.

Keywords: Privacy · PII · LLM · Conversational agents · AI-driven tools · Data detection

1 Introduction

In today's digital age, the collection and aggregation of personal data by various entities can generate detailed profiles of individuals, raising significant concerns regarding privacy and control over personal information [10]. This issue is especially pronounced in the context of AI-driven tools, which often require access to vast amounts of user input to function properly [13]. These systems may inadvertently collect sensitive data such as names, locations or phone numbers through natural language interactions and sometimes without the user being aware of the extent of the data that is being gathered and its potential uses. As these tools become more embedded in daily life [6], the potential of misuse, unintended data sharing or profiling increases, highlighting the need for data protection mechanisms and practices in the design and deployment of potential systems [14]. One particular example of AI-driven tools that may compromise privacy are LLM-based conversational agents such as ChatGPT. As shown in

E. Corchado et al. (Eds.): CISIS 2025, CCIS 2807, pp. 91–100, 2026.
https://doi.org/10.1007/978-3-032-19770-2_9

[15], users disclose a lot of personal information to such agents. It is therefore paramount to protect users' privacy in their interaction with those agents.

With the long-term aim of proposing methods to protect users' privacy when interacting with LLM-based conversational assistants, this paper takes a first step by evaluating various out-of-the-box solutions for personal data detection in human conversations with LLM-based agents. We particularly compare two distinct approaches: a traditional rule-based method in Microsoft Presidio and a more modern strategy leveraging small LLMs as data detection agents. Our ultimate goal is to develop a low-latency system with modest hardware requirements, leveraging small LLMs for general PII detection. This article constitutes our initial step in that direction.

To guide our study, we address the following research questions:

- **RQ1:** How do LLMs compare to traditional techniques in PII detection?
- **RQ2:** What are the trade-offs between detection accuracy and computational efficiency?

This paper contributes: (1) a comparative evaluation of PII detection using a rule-based system and three small LLMs, and (2) an analysis of the trade-off between performance and efficiency. The paper is further organized as follows: Sect. 2 reviews related work; Sect. 3 outlines key concepts; Sect. 4 describes our methodology; and Sect. 5 concludes and discusses future work.

2 Related Work

The threat of users disclosing their own or others' personal data while using LLM-based conversational assistants is a growing concern in the field of natural language processing and privacy protection. Recent research [15] has demonstrated that users frequently reveal various types of personally identifiable information (PII) during interactions, including names, emails, passwords, financial details, health-related information, legal conditions, and sexual orientation. This pattern of disclosure highlights the critical need for effective and reliable detection and protection mechanisms tailored for conversational settings.

Several recent solutions have proposed the use of pre-trained or custom large language models (LLMs) for detecting and preventing the inadvertent disclosure of personal data during conversations with LLM-based agents [1,2,4,14]. Many of these approaches employ Microsoft Presidio as a baseline detection tool or incorporate it into the data labeling process [1,2,14,15], leveraging a combination of rule-based and machine learning techniques to improve detection accuracy.

However, the majority of these existing approaches rely on large LLMs (10B or more), which require significant computational resources and incur substantial inference times. This limits their applicability in real-time or resource-constrained environments, where low latency and efficiency are paramount [5]. Beyond technical challenges, ethical concerns emerge from both false negatives—which risk exposing sensitive user data—and false positives, which may lead to unnecessary censorship and reduce user trust [10]. Despite these critical considerations, prior research often emphasizes detection accuracy without fully addressing the balance between computational efficiency and ethical safeguards. In contrast, our research focuses on utilizing smaller LLMs to evaluate their usability as PII detectors while comparing with lightweight tools such as Microsoft Presidio.

3 Background

Under the General Data Protection Regulation (GDPR), personal data is defined as "*any information relating to an identified or identifiable natural person ('data subject')*" (Regulation (EU) 2016/679, Art. 4(1)). This includes not only direct identifiers such as names and identification numbers but also indirect identifiers like location data, online identifiers (e.g. IP addresses), or factors specific to an individual's physical, physiological, genetic, mental, economic, cultural, or social identity. An individual is considered identifiable if they can be recognized, directly or indirectly, through these types of information. In general, PII refers to any information that can directly or indirectly identify an individual. While the EU's GDPR defines personal data broadly, other frameworks such as the U.S. NIST definition of PII (NIST SP 800-122, 2010) follows similar principles but with minor differences in terminology and scope. In this article, we will employ two approaches to detect the above defined PIs: Microsoft Presidio and Large Language Models.

3.1 Microsoft Presidio

Microsoft Presidio[1] is an open-source and free tool that is the state-of-the-art in the detection and anonymization of PII. Presidio uses the following techniques: regular expressions (regex), data parsing, checksums, and some specialized models for Named Entity Recognition (NER).

Presidio's regular expressions are defined as parsing patterns designed to match the structure of specific types of data. Another parsing method uses whitelists and blacklists. This method compares the presence of a certain word in a text to a list to either let it pass (whitelist) or block it (blacklist). In addition, certain structured data incorporate built-in verification mechanisms into the data generation designed to detect forged data through the application of checksums. These mechanisms can also be employed to identify and accurately classify such formatted data.

Presidio also uses NER. This is a Natural Language Processing (NLP) subtask that identifies meaningful entities in a text through a 3-step process (Processing, Template Matching and Entity Sorting). A NER can recognize the following entities: names (of individuals and organizations), locations, dates, phone numbers among a few other types.

The above methods are known to suffer from two common flaws, data mutability, referring to the multiple ways in which the same PII can be expressed,

[1] https://microsoft.github.io/presidio/.

and data similarity, in which different data entities have similar representations [7,11,12].

3.2 Large Language Models (LLMs)

LLMs are Neural Networks deep-trained with large sets of data. A particular model is characterized by its number of parameters (e.g. 1 billion parameters or 1B), architecture (e.g. GPT, BERT or T5, among others), modal (e.g. Text-Only, Image-Only or Multi-modal among others) and training dataset. Due to their size, LLMs usually require a certain amount of Hardware Specs. These LLM can be deployed to perform a variety of tasks, such as data generation, document parsing or even data detection all depending on a prompt (a structured input, like a text, a question, an instruction or example) that the LLM receives and processes. Most Open-Source LLMs are given as a general base model that can be specialized or expanded through additional training.

While theoretically able to solve the data mutability issue, a new one appears in the form of either prompt mutability, as a lesser mutation on a prompt can alter its functionality [8,9]. Also, not every task may be achieved with every LLM model as these may present some hard-coded safeguards, protection mechanisms (like behavioral hard-coding or ethical alignment) to refuse or prevent to do controversial tasks [3].

4 Method

We evaluate different approaches for detecting personally identifiable information (PII) in a conversation with LLM-based Conversational Agents. This evaluation is conducted using Microsoft Presidio and a selection of different LLMs as detection agents. For the evaluation, we use an existing and publicly available dataset of user conversations with LLM-based assistants (as we detail below). Using the dataset, we compare the precision, recall, and F1 of Presidio and the LLMs to detect whether and what type of PII is disclosed in the conversations.

Regarding the LLMs evaluated, three small LLM models ranging from 1B to 10B parameters, were selected: nuExtract v1.5 3.8B[2], a data extraction-specialized model based on Microsoft Phi 3.5 Mini Instruct, Qwen2.5 3B Instruct[3] and Llama 3.2 3B Instruct[4] both of which are general language models.

The idea behind using small LLMs is that, as hardware processing power increases and LLM efficiency improves, it may eventually become possible to deploy them on almost any device (e.g., a smartphone or personal computer) that users can employ to interact with AI-driven applications.

[2] https://huggingface.co/numind/NuExtract-1.5.
[3] https://huggingface.co/Qwen/Qwen2.5-3B-Instruct.
[4] https://huggingface.co/meta-llama/Llama-3.2-3B-Instruct.

4.1 Dataset

For this experiment, we evaluate user prompts of a labeled subset of the ShareGPT52k, a collection of 52,000 conversations with ChatGPT from various users [15]. To compare the performance of the tools with the ground truth, we asked the authors of [15] to share the labels they created with us, and found that not every sample was on the current public available ShareGPT52k repositories, 156 of said labelled conversations could be found on ShareGPT90k[5], an expansion of ShareGPT52k. A total of 4,047 user prompts pertaining to the 156 conversations were used for our experiments. Table 1 shows the labels in the dataset and the number of conversations with that specific label.

Table 1. Dataset labels and number of conversations with each label. Each conversation may have several prompts.

Label	DATE_TIME	EMAIL	LOCATION	NRP	PASSPORT	PERSON	PHONE	URL
Number of Conv.	32	7	29	13	1	20	7	8

While these labels are mostly self-contained, there are two points to consider: NRP is a label defined on Presidio as any information related to Nationality, Religion and Political Group. Also, on recent Presidio versions, the label PASSPORT has been deprecated and split instead to other labels (e.g. US_PASSPORT or IT_PASSPORT).

4.2 Experimental Setting

For Presidio, we used the default SpaCy NER model "en_core_web_lg", keeping all parameters at their default settings.

For the LLMs, a prompt containing: (1) a *basic directive* (e.g., "You are a PII detection model... The required JSON fields are:") and a *template*, (2) a structured format specifying the desired data to extract (e.g., "Name": Person's full name.\n - "Birth_Date": Date of birthSPSVERBc6 - "Age": Age of the personSPSVERBc6...) and (3) an example of the template was sent, followed by the text to analyze. No additional training was performed on any model. Outputs were post-processed to remove undesired results (e.g., "No data found" responses, malformed or hallucinated templates) before computing performance metrics and timing.

The LLMs were deployed on a SLURM instance of an HPC cluster equipped with one Nvidia A40 GPU and one logical CPU of an AMD EPYC 7453, running a Miniconda Python environment on Ubuntu with vLLM serving the target models.

[5] https://huggingface.co/datasets/RyokoAI/ShareGPT52K.

4.3 Metrics

Detection Performance. The performance of detection can be measured using two main metrics: Precision and Recall. These metrics indicate how accurate and reliable an approach is, and can be combined into the F1 score to provide a comprehensive assessment of performance. We use all three metrics to evaluate the performance of each labeled data type in the dataset for every approach (Presidio and LLMs). The results will also be reported in an aggregated form, employing both micro and macro averaging due to the dataset being highly unbalanced and not defining if certain data types have more weight than others in our context.

For micro averaging:

$$fp_\mu = \sum_{i=0}^{n} fp_a \quad tp_\mu = \sum_{i=0}^{n} tp_a \quad fn_\mu = \sum_{i=0}^{n} fn_a \tag{1}$$

Being fp_a, tp_a and fn_a, the false positives, true positives and false negatives of an individual label. We then will use these new values fp_μ, tp_μ and fn_μ to re-calculate Precision, Recall and F1 score.

For macro averaging:

$$Metric_M = \frac{1}{N} \sum_{i=0}^{n} Metric_a \tag{2}$$

where $Metric_a$ is the Precision, Recall or F1 score of the individual labels.

Processing Time. Beyond detection performance, measuring the computational cost of each method is crucial. We assess this by the average processing time each method requires to produce a result. In our case, we use:

$$t_{prompt} = \frac{t_{total}}{N_{cases}} \tag{3}$$

where t_{prompt} is the average computing time per prompt, t_{total} is the total computing time a test has taken to parse all prompts, and N_{cases} is the amount of processed prompts. Max and min times per model will also be presented.

5 Results

5.1 PII Detection Performance Results

Table 2 shows the detection performance results for Presidio. The tool achieves perfect recall (Recall = 1), successfully identifying almost all instances of relevant PII. However, precision remains very low, indicating a high number of false positives. A notable exception is the `PASSPORT` label, where the single instance is incorrectly classified as a U.S. driver's license (`US_DRIVER_LICENCE`). This outcome is expected, given Presidio's rule-based nature, which relies heavily on pattern matching and may conflate structurally similar entities.

Table 2. Presidio performance metrics.

	DATE_TIME	EMAIL	LOCATION	NRP	PASSPORT	PERSON	PHONE	URL
Precision	0.23	0.47	0.22	0.11	0.00	0.13	0.21	0.09
Recall	1.00	1.00	1.00	1.00	0.00	1.00	1.00	1.00
F1	0.47	0.28	0.31	0.26	0.00	0.23	0.29	0.08

Table 3. nuExtract performance metrics.

	DATE_ TIME	EMAIL	LOCATION	NRP	PASSPORT	PERSON	PHONE	URL
Precision	0.00	0.50	0.18	0.29	0.00	0.23	1.00	0.00
Recall	0.00	0.14	1.00	0.15	0.00	0.35	0.14	0.00
F1	0.00	0.22	0.31	0.20	0.00	0.27	0.25	0.00

Table 3 presents the results obtained with nuExtract. Overall, both precision and recall are relatively low. Using the provided prompt template, the model failed to identify relevant instances for the PASSPORT, URL, and DATE_TIME labels.

Table 4. Qwen 2.5 performance metrics.

	DATE_ TIME	EMAIL	LOCATION	NRP	PASSPORT	PERSON	PHONE	URL
Precision	0.38	0.38	0.18	0.00	0.00	0.56	0.11	0.00
Recall	0.09	0.71	1.00	0.00	0.00	0.25	0.14	0.00
F1	0.15	0.50	0.31	0.00	0.00	0.34	0.12	0.00

The results obtained with Qwen, shown in Table 4, are broadly comparable to those of nuExtract. For this particular model, we did not observe any extracted data labeled as URL, PASSPORT, or NRP. This absence does not necessarily imply that such data were not detected or present in the input; rather, it may be due to mislabeling, omission, or incomplete or malformed outputs.

Table 5 presents a scenario similar to that observed with the other two LLMs. In this case, however, the model appears to have missed detecting instances of the NRP, PASSPORT, and URL labels.

Table 6 compares macro- and micro-averaged metrics for all models. Overall performance remains similar across experiments, with minor precision improvements on some labels. Each LLM shows a slight preference for certain data types, with Llama performing somewhat worse—likely due to built-in content filtering. Qwen and nuExtract exhibit comparable results, but differ in which data types are missed (DATE_TIME for nuExtract and NRP for Qwen).

If we compare the number of outputs containing information to those with genuinely significant content on all three models across all the extracted data:

Table 5. Llama 3.2 performance metrics.

	DATE_TIME	EMAIL	LOCATION	NRP	PASSPORT	PERSON	PHONE	URL
Precision	0.38	0.14	0.18	0.00	0.00	0.14	0.14	0.00
Recall	0.16	0.29	1.00	0.00	0.00	0.05	0.29	0.00
F1	0.22	0.19	0.31	0.00	0.00	0.07	0.19	0.00

Table 6. Performance comparison between models.

	$Precision_\mu$	$Precision_M$	$Recall_\mu$	$Recall_M$	$F1_\mu$	$F1_M$
Presidio	0.16	0.18	1.00	0.88	0.28	0.29
nuExtract	0.20	0.27	0.34	0.22	0.25	0.16
Qwen 2.5	0.22	0.20	0.37	0.28	0.27	0.18
Llama 3.2	0.18	0.12	0.33	0.22	0.23	0.12

Llama has found more results than the other two LLMs. This was surprising in early analyses, but it may be an indicator that it is more prone to hallucinate even on a temp-0 setting (highly deterministic), possibly due to completion-driven hallucinations triggered by restricted usage flags by the PI detection intent in the reused prompt (Table 7).

5.2 Processing Time

Table 8 shows a comparison of the minimum and maximum processing time related to a single prompt, the average time per prompt and the total time per full dataset parse.

Presidio is the fastest solution with around 490 s per full parse, corresponding 123 ms per prompt. All LLMs were at least an order of magnitude slower than Presidio, having Qwen as the fastest LLM with around 9 s per prompt, followed by Llama with 10 s per prompt and far behind nuExtract with 24 s per prompt. A notable observation is that Qwen and Llama consistently process prompts in less than half the time required by nuExtract. Both models show minimum and maximum processing times significantly lower than nuExtract, with minimum times comparable to Presidio. The occasional high maximum times for these LLMs likely result from outliers caused by hallucinations on large prompts.

Table 7. Extracted data comparison between LLMs.

Model	Significant Outputs	Total Outputs
nuExtract	200	825
Qwen	197	753
Llama	209	1627

Table 8. Computation time comparison.

	t_{min} (s)	t_{max} (s)	t_{total} (s)	t_{prompt} (s)
Presidio	0.01	89	490	0.12
nuExtract	23.74	585	85000	24.00
Qwen 2.5	0.53	114	35505	9.00
Llama 3.2	0.14	185	36444	10.00

6 Conclusions and Future Work

At first glance, the use of LLMs does not appear to offer a clear advantage over traditional techniques. The small LLMs tested yielded results comparable to those of Presidio, with two of the three models slightly outperforming it in terms of precision—albeit at the cost of significantly lower recall and slower processing times (RQ2). Nevertheless, we observed that LLMs have the potential to recognize and categorize a broader range of personal data types than rule-based tools like Presidio, and may offer a more flexible and interpretable approach to structuring the extracted information (RQ1). It is also worth noting that using prompts tailored to specific categories of personal data may help improve output stability, potentially enhancing alignment with real-world deployment requirements.

Out-of-the-box tools, whether small LLMs or tools like Presidio, may lead to potential real systems able to at least partially anonymize user prompts to other LLM-based Conversational Agents like ChatGPT in a way to safeguard user data and privacy. Yet, performance seems to be the limiting factor, as false positives risk censoring harmless content and reducing trust. Conversely, false negatives pose a privacy risk by allowing sensitive data to go undetected. While LLMs offer strong performance in nuanced cases, their inference time and resource requirements pose challenges for low-latency applications. In contrast, lightweight tools like Microsoft Presidio offer high-speed processing but lower precision in complex contexts. These findings suggest that neither rule-based systems nor small LLMs alone as out-of-the-box tools may be sufficient for robust PII detection in real-world settings. However, their complementary strengths open the door for hybrid solutions. Our work provides a foundation for such systems by quantifying trade-offs and feasibility. This is essential for deploying privacy-preserving conversational agents in practical, latency-sensitive environments.

Considering these results, our future work is oriented toward the development of a hybrid approach that combines the high recall and low latency of Microsoft Presidio as a first filtering step, reducing the input size for LLMs and thereby improving scalability, followed by targeted inspection by LLMs. This strategy aims to balance efficiency with the nuanced reasoning capabilities of LLMs. Moreover, as our current study is limited to a single dataset and three small LLMs, future work will expand the evaluation to include additional datasets, model families, and model sizes to better assess the generalizability and robustness of our findings. We also plan to gain further insight into the causes of variability in model performance through a qualitative analysis of outputs.

Acknowledgements. This work is partially supported and funded by Spanish Government project PID2023-151536OB-I00 and by the INCIBE's strategic SPRINT (Seguridad y Privacidad en Sistemas con Inteligencia Artificial) C063/23 project with funds from the EU-NextGenerationEU through the Spanish government's Plan de Recuperación, Transformación y Resiliencia.

References

1. Asthana, S., Mahindru, R., Zhang, B., Sanz, J.: Adaptive PII mitigation framework for large language models. arXiv preprint arXiv:2501.12465 (2025)
2. Asthana, S., et al.: Deploying privacy guardrails for LLMs: a comparative analysis of real-world applications. arXiv preprint arXiv:2501.12456 (2025)
3. Bai, Y., et al.: Constitutional AI: harmlessness from ai feedback. arXiv preprint arXiv:2212.08073 (2022)
4. Böhlin, F.: Detection & anonymization of sensitive information in text: AI-driven solution for anonymization (2024)
5. Chandran, R., Tan, M.-L.: Efficiently scaling LLMs challenges and solutions in distributed architectures. Baltic Multidisciplinary Res. Lett. J. **2**(1), 57–66 (2025)
6. Chen, T., Gascó-Hernandez, M., Esteve, M.: The adoption and implementation of artificial intelligence chatbots in public organizations: evidence from us state governments. American Rev. Public Administration **54**(3), 255–270 (2024)
7. Gambarelli, G., Gangemi, A., Tripodi, R.: Is your model sensitive? Spedac: a new benchmark for detecting and classifying sensitive personal data. arXiv preprint arXiv:2208.06216 (2022)
8. Salinas, A., Morstatter, F.: The butterfly effect of altering prompts: how small changes and jailbreaks affect large language model performance. arXiv preprint arXiv:2401.03729 (2024)
9. Sclar, M., Choi, Y., Tsvetkov, Y., Suhr, A.: Quantifying language models' sensitivity to spurious features in prompt design or: how i learned to start worrying about prompt formatting. arXiv preprint arXiv:2310.11324 (2023)
10. Solove, D.J.: Understanding privacy. Harvard university press (2010)
11. Truong, A., Walters, A., Goodsitt, J.: Sensitive data detection with high-throughput neural network models for financial institutions. arXiv preprint arXiv:2012.09597 (2020)
12. Tutuianu, A., et al.:. Efficient statistical techniques for detecting sensitive data (2023). US Patent 11,599,667
13. Wei, J., Kim, S., Jung, H., Kim, Y.-H.: Leveraging large language models to power chatbots for collecting user self-reported data. In: Proceedings of the ACM on Human-Computer Interaction, 8(CSCW1):1–35, April 2024. ISSN 2573-0142. https://doi.org/10.1145/3637364
14. Yang, J., Zhang, X., Liang, K., Liu, Y.: Exploring the application of large language models in detecting and protecting personally identifiable information in archival data: a comprehensive study. In: 2023 IEEE International Conference on Big Data (BigData), pp. 2116–2123. IEEE (2023)
15. Zhang, Z., et al.: It's a fair game, or is it? Examining how users navigate disclosure risks and benefits when using LLM-based conversational agents. In: Proceedings of the 2024 CHI Conference on Human Factors in Computing Systems, pp. 1–26 (2024)

Adaptive Identity Token from User Attributes for Authentication Through Controlled Execution Environment

Shashank Tripathi(✉), Kai Hendrik Wöhnert, and Volker Skwarek

RTC CyberSec, Hamburg University of Applied Sciences, Berliner Tor 5, 20099 Hamburg, Germany
shashank.tripathi@haw-hamburg.de
https://www.haw-hamburg.de/hochschule/life-sciences/forschung/ftz-cybersec

Abstract. With an increased emphasis on software and applications, they are becoming more secure through built-in malware recognition. It has prompted adversaries to increasingly target identity infrastructures, thereby contributing to a pronounced global surge in identity-based attacks. This paper addresses the problem by introducing *IdentiToken*, an authentication framework that derives tokens from a structured set of user and environmental attributes. Attributes are grouped into different classes, each weighted to reflect its sensitivity to change. IdentiToken supports similarity-based validation, enabling partial matches to be interpreted meaningfully rather than reduced to binary outcomes. We analyse the system's behaviour under adversarial scenarios and evaluate its sensitivity to attribute changes in a controlled environment. Although the approach is still exploratory in terms of deployment, our results suggest that structured, attribute-derived tokens may provide a useful foundation for developing more flexible and context-aware authentication mechanisms.

Keywords: identity-based authentication · tokenization · attribute-based identification

1 Introduction

Most identity systems authenticate users via static secrets like passwords, cryptographic keys or biometrics, which are prone to theft and reuse. In 2024, identity-based attacks surged by 71%, with 80% linked to credential misuse [1]. Breaches like the Okta incident [2] and flaws in certificate chains [3] underscore limitations of established credential-based authentication.

However, the emphasis on verifying knowledge of a secret makes such systems vulnerable to replay attacks, where an adversary reuses previously captured valid credentials to impersonate a legitimate user. Moreover, binary verification neglects the context or additional user-related environment in which the authentication occurs. Verification of such additional information adds a layer to verify

E. Corchado et al. (Eds.): CISIS 2025, CCIS 2807, pp. 101–113, 2026.
https://doi.org/10.1007/978-3-032-19770-2_10

a user with higher confidence. While MFA [4] and decentralised identity (DID) frameworks [5,6] enhance user control, they retain dependency on bearer credentials and have been prone to attack. Addressing the current limitations:
Can we design a privacy-preserving, behaviour-sensitive token that tolerates drift in identity while remaining verifiable?

The solution proposed in this paper addresses this by generating identity tokens from weighted attributes and measuring similarity across sessions, enabling flexible and context-aware authentication in dynamic or semi-trusted environments (e.g., IoT, federated identity, edge computing).

Our Contributions:

- Introduced a structured identity model using attribute segmentation for tracking individual attribute contributions and identity drift.
- Proposed a cryptographic token generation algorithm that preserves entropy during hash truncation and achieves cross-attribute diffusion using discrete reduction windows.
- Implemented the system in a controlled and portable runtime environment to ensure reliable attribute recording, tested with realistic changes (e.g., keystroke patterns, geolocation shifts).
- Conducted formal analysis demonstrating resistance to replay attacks and tampering.

2 State of the Art

ISO/IEC 24760-1:2019 defines identity as a set of attributes associated with an entity and authentication as the process of verifying that identity by comparing presented attributes against previously stored or registered values. Static-password systems remain widespread but are easily compromised [7]. Widely used through protocols like OAuth 2.0 and OpenID Connect [8], tokens such as JSON web tokens (JWTs) enable federated identity but suffer from vulnerabilities like token substitution and replay attacks. WebAuthn [9], part of the FIDO2 standard, uses public-key pairs for passwordless login.

WebAuthn, part of the FIDO2 standard, uses public-key pairs for passwordless login. While offering phishing resistance, real-world issues like stale challenge reuse [3], misconfigured verifiers and key management remain [10].

Macaroons [11] allow constrained delegation allowing it to be passed from one party to another with added restrictions (e.g., limiting time, endpoint, user role) but are bearer-based, meaning, whoever possesses them can use them. If leaked or intercepted, they can be misused.

Attribute-Based authentication frameworks [12] use cryptographically signed claims (e.g., age, role). However, traditional models raise privacy concerns due to semantic linkability and limited revocation mechanisms.

Our approach differs by integrating multiple attributes into a unified identity token, enabling similarity-aware authentication and session drift tolerance. Unlike traditional systems, it supports partial identity acceptance and adaptive security logic.

3 Tokenization of Identity

This section describes how identity attributes are transformed into a cryptographic token using a modified variant of the FaRHash [13] algorithm. The token must react differently to the type of attributes, formally termed as static, dynamic and volatile. Static attributes are the most stable components of an identity, remaining consistent across an entity's lifecycle and providing foundational trust, for example, a vehicle's chassis number or a server's hardware serial number. Dynamic attributes vary over time due to operational context or usage patterns but still contribute meaningfully to identity, for instance, the average operating speed of an industrial machine or the access patterns of a database over time. Volatile attributes, like room temperature in a data centre or tempo rary file usage pattern on a system change, often have little impact on identity when viewed alone. However, when many such changes occur together, they can alter the system's behaviour and affect its overall identity. By cryptographically processing these attributes, we generate a token called *IdentiToken*.

Let the attribute space of an entity be partitioned into three disjoint subsets:

- Static attributes: $\mathcal{S} = \{s_1, s_2, \ldots, s_p\}$
- Dynamic attributes: $\mathcal{D} = \{d_1, d_2, \ldots, d_q\}$
- Volatile attributes: $\mathcal{V} = \{v_1, v_2, \ldots, v_r\}$

Each group contributes differently to the token structure. Static attributes influence the token globally by acting as a seed, while dynamic and volatile attributes contribute direct entropy.

3.1 Token Generation

Step 1: Seed Construction. We begin by creating a seed from the static attributes. Each static attribute s_j is first encoded to binary, and the seed is formed by concatenating all such encodings:

$$\mathsf{SEED} = \|_{j=1}^{p} \mathsf{Bin}(s_j) \tag{1}$$

Any change in static attributes yields a completely new seed, and therefore, a drastically different token.

Step 2: Attribute-Specific Hashing. Next, for every $x_i \in \mathcal{D} \cup \mathcal{V}$, we concatenate its binary encoding with the seed and hash the result:

$$x_i' := \mathsf{Bin}(x_i) \| \mathsf{SEED}, \qquad h_i := \mathsf{H}(x_i')$$

This ensures that attribute values are kept private while still contributing entropy to the token.

Step 3: Truncation by Attribute Type. To control each attribute's influence, hash outputs are truncated differently:

$$\tilde{h}_i := \begin{cases} \mathsf{Slice}_n(h_i), & \text{if } x_i \in \mathcal{D} \\ \mathsf{Slice}_m(h_i), & \text{if } x_i \in \mathcal{V} \end{cases}$$

Here, $n > m$ ensures that dynamic attributes carry more weight in the final token than volatile ones. We also choose $n \bmod m \neq 0$ to promote diffusion in the next step.

Step 4: Stream Assembly. All truncated hashes are concatenated in a predefined order π:

$$\mathsf{H\_S} := \|_{x_i \in \pi(\mathcal{D} \cup \mathcal{V})} \tilde{h}_i$$

Step 5: Cross-Attribute Diffusion. The stream $\mathsf{H\_S}$ is divided into fixed-size windows of length $\ell = m$:

$$W_i := \mathsf{H\_S}[\ell \cdot i : \ell \cdot (i+1)] \tag{2}$$

If the last chunk is shorter than ℓ, we pad it with XOR-neutral hex digits 0.

Each window is reduced to one hex digit by XORing all its bytes:

$$\mathsf{T}_i := \bigoplus_{j=0}^{\ell-1} W_i[j]$$

Finally, the IdentiToken is formed as (Fig. 1):

$$\mathsf{T} := \|_i \mathsf{T}_i$$

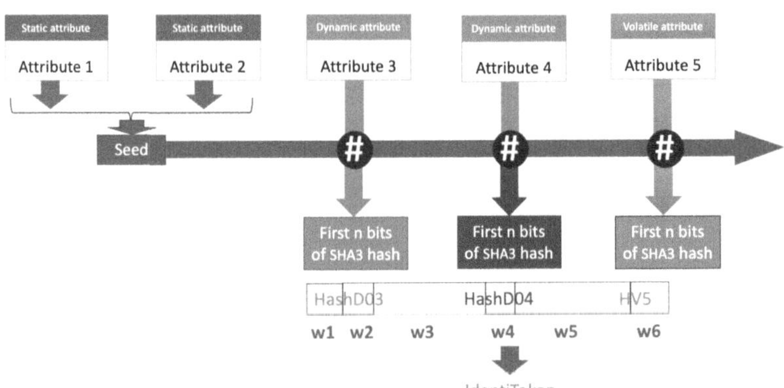

Fig. 1. Token construction from weighted attributes.

Compact Formalization.

$$\mathsf{T} = \|_{i=0}^{w-1} \bigoplus_{k=0}^{\ell-1} \left[\mathsf{H\_S} \left(\|_{x \in \pi(\mathcal{D} \cup \mathcal{V})} \mathsf{Slice}_{b_x} \left(\mathsf{H} \left(\mathsf{Bin}(x) \| \mathsf{SEED} \right) \right) \right) \right]_{i \cdot \ell + k}$$

Algorithm. The steps of the algorithm have been implemented with the pseudocode below.

Algorithm 1 IdentiToken Generation

Require: $\mathcal{S}$, $\mathcal{D}$, $\mathcal{V}$, parameters n, m
Ensure: Token T
$SEED \leftarrow \|_{s \in \mathcal{S}} \mathsf{Bin}(s)$
$H\_S \leftarrow$ empty string
for $x \in \mathcal{D} \cup \mathcal{V}$ **do**
  $x' \leftarrow \mathsf{Bin}(x) \| SEED$
  $h \leftarrow \mathsf{H}(x')$
  Append $\mathsf{Slice}_n(h)$ or $\mathsf{Slice}_m(h)$ to H_S
end for
if $|H\_S| \bmod m \neq 0$ **then**
  Pad H_S with 0s
end if
for each window W_i in H_S **do**
  $\mathsf{T}_i \leftarrow \bigoplus W_i$
  Append T_i to T
end for
return T

3.2 Formal Proof of the Design Properties

Each identity attribute is hashed independently after being concatenated with a seed derived from static attributes (Eq. 1). Dynamic and volatile attributes are truncated to n and m bytes, respectively, with $n > m > 16$. Choosing $n \bmod m \neq 0$ ensures that window boundaries misalign with segment boundaries, promoting inter-attribute diffusion.

The resulting hash stream is divided into non-overlapping windows of size $\ell = m$, each reduced to one byte via XOR. While XOR is linear, applying it across randomised truncated slices preserves entropy under the random oracle model. Given $n \geq 128$ and $m \geq 28$, each token window reflects high-entropy mixing and retains cryptographic properties such as pre-image and collision resistance.

Static Sensitivity. The seed–derived from static attributes–affects every attribute hash. A change in any $s_j \in \mathcal{S}$ causes full regeneration of the hash stream. If the total stream length is $L = d \cdot n + v \cdot m$, the number of windows is $w = \lfloor \frac{L}{\ell} \rfloor$. Under SHA-3's avalanche property, token similarity drops sharply. The probability of a collision between two tokens due to seed change is bounded by:

$$\Pr[\mathsf{T}(\mathsf{SEED}) = \mathsf{T}(\mathsf{SEED}')] \leq \frac{1}{2^{32w}}$$

For $w = 12$, this yields $\approx 2^{-384}$, a negligible probability.

Dynamic Proportionality. Let α be the fraction of modified dynamic attributes. Each contributes n bytes, affecting approximately $\frac{n}{\ell}$ token windows. Thus, the expected token difference due to partial changes in $\mathcal{D}$ is:

$$\Pr[\mathsf{T} \neq \mathsf{T}'] \approx \alpha \cdot \frac{n}{\ell}$$

This satisfies proportional sensitivity to behavioural drift.

Volatile Tolerance. Volatile attributes fluctuate frequently. To tolerate minor drift, the similarity is computed over the volatile segments of two tokens. Let $d_{\mathcal{V}}(\cdot, \cdot)$ be a Hamming distance restricted to volatile-derived parts:

$$d_{\mathcal{V}}(\mathsf{T}_1^{(v)}, \mathsf{T}_2^{(v)}) \leq \delta \quad \Rightarrow \quad \text{identity preserved}$$

If this threshold is exceeded, identity is re-evaluated.

Privacy Preservation. Only 4 bytes per window are visible. Given SHA-3's 256-bit output and token length of $4w$ bytes, the probability of inverting an attribute is:

$$\Pr[\mathcal{A}(\mathsf{T}) = x_i] \leq \frac{1}{2^{224}}$$

Due to misaligned segments and XOR blending, one token byte never maps cleanly to one attribute.

Pre-image Resistance. Each token byte is derived from multiple hash slices. Given visibility of 4 bytes per window and $\lambda = 256$, pre-image resistance is bounded by:

$$\Pr\left[\text{find } x_i \mid \mathsf{T}\right] \leq \frac{1}{2^{224}}$$

This bound follows from the fact that revealing 32 out of 256 bits of a cryptographically secure hash leaves 224 bits hidden, assuming no structure leaks and each attribute influences misaligned, XOR-blended segments. As a result, an adversary observing a full window gains only partial, non-direct information about any x_i.

Collision Resistance: Applying the birthday bound to this L-bit token, the probability of a collision among Q such tokens is:

$$\Pr[\text{collision among } Q \text{ tokens}] \leq \frac{Q^2}{2^L}$$

With $L \geq 384$ or more, the collision probability remains negligible for practical values of Q.

3.3 Similarity Score of Tokens

Since authentication is the process of verifying identity, we propose a model that uses IdentiToken as the basis for this verification. Authentication is performed by comparing a freshly generated token with a previously stored reference, enabling a lightweight yet robust comparison mechanism. For clarity and focus, we abstract away the complexities of full authentication frameworks and concentrate on this core verification step. In this simplified model, tokens are represented as hex-encoded bitstrings. To quantify the degree of change between two tokens, we define a similarity score using the normalised Hamming distance:

$$\text{Sim}(\mathsf{T}_1, \mathsf{T}_2) = 1 - \frac{H(\mathsf{T}_1, \mathsf{T}_2)}{|\mathsf{T}_1|}, \tag{3}$$

where $H(*, *)$ is the Hamming distance and $|\mathsf{T}_1|$ is token length in hex representation. A score of 1 implies an exact match; 0 denotes total difference.

Threshold Selection. The thresholds, τ_d for dynamic similarity and δ for volatile similarity, are not fixed algorithmic constants but are instead chosen based on application requirements. In highly critical scenarios such as financial access control, a stricter threshold (e.g., $\tau_d \geq 0.9$) enforces continuity with minimal tolerance. In contrast, less sensitive use-cases like user personalisation or adaptive content delivery can operate reliably with a relaxed bound (e.g., $\tau_d \geq 0.6$). The volatile threshold δ is likewise configured according to the volatility profile acceptable within the specific context.

Dynamic Changes. Compare tokens T_1 and T_2, differing in 4 of 5 dynamic attributes. Static and volatile attributes remain unchanged.

```
T_1 = 8f8e...c51053c5a861fa89571240...fd4a363
T_2 = 96c2...7e7e7d5a861fa89571240...fd4a363
```

$$|\mathsf{T}| = 86, \quad H(\mathsf{T}_1, \mathsf{T}_2) = 46 \quad \Rightarrow \quad \text{Sim} = 1 - \frac{46}{86} = 0.4651$$

This indicates substantial identity drift. For instance, in moderately strict scenarios where limited drift is acceptable (e.g., $\tau_d = 0.75$), access would be denied. However, lower similarity scores may still permit degraded or provisional access depending on policy. For example, similarity in the range $0.5 < \text{Sim} < 0.75$ might trigger secondary checks, fallback authentication or limited access modes. As established earlier, τ_d and δ are not inherent to the algorithm itself; instead, they are configurable parameters determined by the specific needs and sensitivity of the deployment context.

Next, T_3 differs from T_1 in only 1 dynamic attribute:
`T_3 = 8f8e...0132b4933d3f...c51053c5a861fa...fd4a363`

$$H(\mathsf{T}_1, \mathsf{T}_3) = 10 \quad \Rightarrow \quad \text{Sim} = 1 - \frac{10}{86} = 0.8837$$

This score supports identity continuity for $\tau_d < 0.88$.

Volatile Drift. Now T_4 differs from T_1 in 5 of 6 volatile attributes:
`T_4 = 8f8e...c51053c4b4df0d...2aeb55a363`

$$H(\mathsf{T}_1, \mathsf{T}_4) = 17 \quad \Rightarrow \quad \text{Sim} = 1 - \frac{17}{86} = 0.8023$$

Extract volatile parts (26 hex digits):

$$\mathsf{T}_1^{(v)} = \texttt{24098fdd0f09559364cfd4a363}$$
$$\mathsf{T}_4^{(v)} = \texttt{c4b4df0d0ae7e8132aeb55a363}$$

$$H(\cdot) = 17 \quad \Rightarrow \quad \text{Sim}_v = 1 - \frac{17}{26} = 0.3462$$

A drop in volatile similarity below $\delta = 0.4$ indicates significant context change. Even with unchanged dynamics, re-authentication may be required.

Static Change. T_5 differs from T_1 only in one static attribute, affecting the seed:
`T_5 = e3ea...7afc...b14cafcdce...aeb55`

$$H(\mathsf{T}_1, \mathsf{T}_5) = 78 \quad \Rightarrow \quad \text{Sim} = 1 - \frac{78}{86} = 0.0930$$

Despite no dynamic or volatile changes, a minimal static modification regenerates the entire token, demonstrating seed centrality.
These cases show how token similarity captures granular identity shifts. The next section embeds this logic within secure local execution environments.

4 Demonstration

We now instantiate IdentiToken in a controlled test case to demonstrate its behaviour under identity drift. Consider a user who installs a benign but critical application (e.g., a trading terminal). While functionally irrelevant, the installation may modify the environment. This allows us to examine IdentiToken's proportionality and responsiveness.

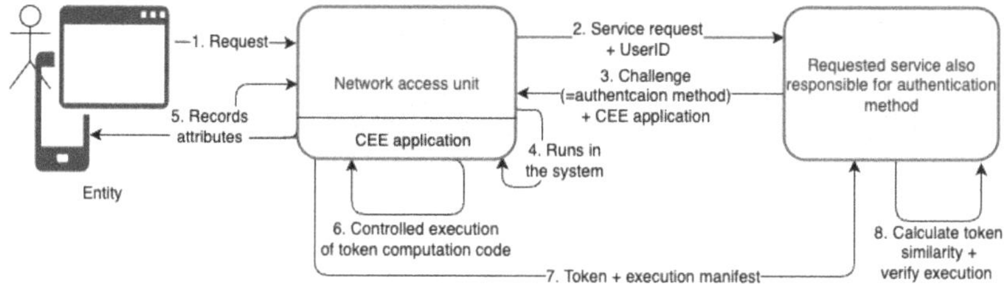

Fig. 2. Component view of IdentiToken computation and verification using a CEE.

All authentication logic was executed inside a Docker based controlled execution environment (CEE), following the architecture in Fig. 2. The implementation[1] consisted of three stages: (i) local attribute acquisition, (ii) token generation, and (iii) server submission. Attributes were acquired from within the container, using Python.

Attribute selection within each category is use case dependent and based on relevance. An attribute considered static in one context may be dynamic in another or even excluded entirely from identity calculation in a third scenario.

Static Attributes: This category includes properties with long-term stability that significantly contribute to an entity's identity. Selected attributes include MAC address, memory configuration and CPU characteristics. As a behavioural attribute, keystroke dynamics were used [14]. In the implementation, the user types the phrase "the lazy fox jumps over the brown dog", consisting of all the English alphabet five times. Keystrokes with durations slower than the session's Gaussian mean were retained, as described in [15].

Dynamic Attributes: These attributes exhibit moderate variability within bounded ranges and play a partial role in shaping the entity's identity. For demonstration purposes, we selected attributes such as geolocation and IP address, which tend to remain stable within sessions or short windows but may vary across environments.

Volatile Attributes: This category captures rapidly changing properties that reflect transient context. In our setup, we used runtime disk size and network latency, both of which are highly sensitive to immediate environmental conditions.

Tokens (represented in 44 hex digits) were computed and compared using Hamming similarity, with scatter plots of byte-wise Spearman ranks shown in Fig. 3.

Six cases highlight IdentiToken's response: (a) No change: tokens match exactly (Sim = 1.0000). (b) One static change: Sim = 0.043. (c)Two dynamic changes: Sim = 0.6190. (d) One dynamic change: Sim = 0.7619. (e)Two volatile changes: Sim = 0.8809. (f)One volatile change: Sim = 0.9286.

[1] The code is publicly available at https://osf.io/8cepa/?view_only=5d58a8b2e72f4d42a1c47686faf50b29.

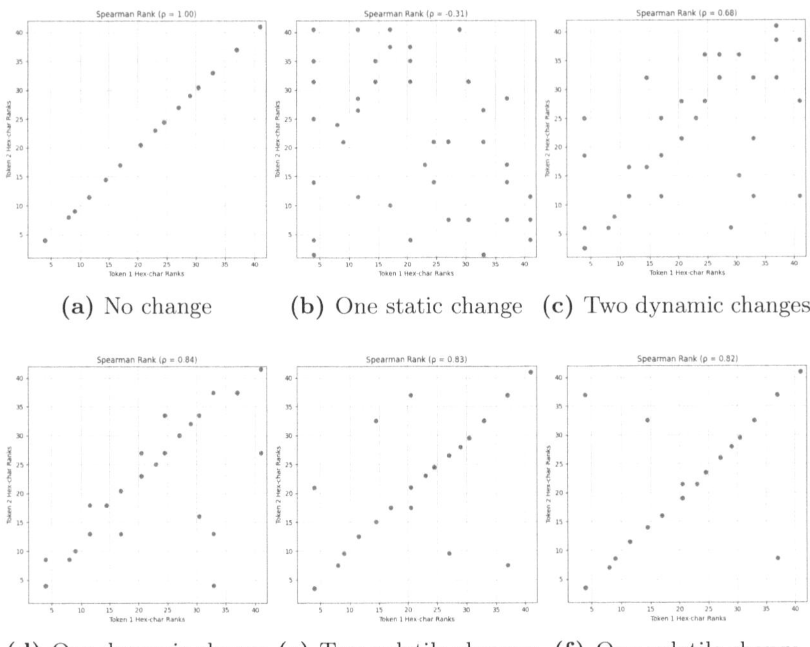

(a) No change **(b)** One static change **(c)** Two dynamic changes

(d) One dynamic change **(e)** Two volatile changes **(f)** One volatile change

Fig. 3. Spearman Rank Scatter Plots Comparing Token Variations.

The experiment demonstrated that IdentiToken is sensitive to static attribute shifts, moderately reactive to dynamic changes and tolerant to small volatile drift. The CEE ensures privacy and determinism in token generation, supporting secure deployment in resource-constrained or privacy-critical settings.

5 Threat Model and Analysis

We assume an adversary capable of intercepting tokens, manipulating the execution environment, and forging inputs. IdentiToken is designed to resist these under three guarantees:

- Cryptographic irreversibility of SHA-3,
- Trusted token generation inside a verifiable CEE,
- Threshold-based acceptance allowing bounded drift.

5.1 Replay Attack

Threat: An intercepted token $\mathsf{T}_u(t)$ is replayed to impersonate a user.

Assumptions: The adversary can sniff tokens but lacks access to the CEE or true attributes.

Defence: Volatile attributes (e.g., IP, timestamp) change naturally. The probability of replay success is:

$$\Pr[\text{replay accepted}] = \prod_{i=1}^{r} p_i$$

$$\Pr[\text{replay rejected}] = 1 - \Pr[\text{replay accepted}]$$

Example: With IP = 0.3, UA = 0.8, DF = 0.6:

$$\Pr = 0.3 \times 0.8 \times 0.6 = 0.144 \Rightarrow \text{Rejection} = 0.856$$

Optional freshness (e.g., TTL) and signed execution manifests can further reduce replay success.

5.2 Token Forgery

Threat: The adversary crafts a forged token $\mathsf{T}_{\mathcal{A}}$ that passes similarity checks.

Assumptions: Partial or full leakage of legitimate attribute values; no CEE access.

Objective: Forge:

$$\mathsf{T}_{\mathcal{A}} = \mathcal{H}(\mathbf{S}_u \oplus \mathbf{a}_s,\ \mathbf{D}_u \oplus \mathbf{a}_d,\ \mathbf{V}_u \oplus \mathbf{a}_v)$$

with:

$$\text{Sim}(\mathsf{T}_{\mathcal{A}}, \mathsf{T}_u^{\text{ref}}) \geq \Theta_{\text{accept}}$$

Defense: Token similarity is weighted: $w_s > w_d > w_v$. Acceptance requires:

$$E[\text{Sim}] = 1 - (w_s\delta_s + w_d\delta_d + w_v\delta_v) \geq \Theta_{\text{accept}}$$

Even small static changes (δ_s) lead to rejection, while full drift in volatile attributes can be tolerated if $w_s\delta_s + w_d\delta_d$ remains within bounds.

6 Discussion

IdentiToken affirms that identity can be adaptively asserted via weighted attributes, similarity thresholds, and controlled execution. Unlike static tokens (OAuth2, OIDC) or fixed-attribute frameworks (SAML, X.509), it recomputes identity based on current behaviour and environment, avoiding persistent secrets and rigid profiles.

Its structure encodes static, dynamic and volatile traits, facilitating graded identity responses. Static components dominate identity, dynamic properties influence the token shape and volatile elements allow benign drift without immediate rejection. Similarity scoring supports nuanced decisions beyond binary matches, distinguishing contextual change from attack.

Compared to MFA [16], which verifies isolated factors sequentially one after the other, but IdentiToken compares them simultaneously. This transfers the identity directly into an n-dimensional feature space. This prevents an attacker from trying out the sensor dispersion for each attribute individually until they hit the overlap area through imitation and dispersion. Instead, all detections are transferred to the feature space and evaluated simultaneously. This makes it exponentially more difficult to exploit sensor scattering in a targeted manner.

Real-time telemetry (e.g., latency, network timing) can be embedded to update identity without user burden. This supports lightweight, context-aware authentication in zero-trust and dynamic settings. Token recomputation in verifiable CEEs eliminates reliance on stored secrets, shifting trust to code and context rather than storage.

The model is especially suited for constrained devices, edge authentication, and continuously evolving environments. However, deployment requires a sandbox on the client, which may limit universality. Broader CEE integration (e.g., native browser/runtime support) may ease this in future applications.

Token validation works on a similarity continuum, enabling partial access, reverification, or rejection based on attribute drift. This allows both interpretability and policy flexibility.

Limitations remain: defining a stable yet discriminative attribute set is specific to the scenario; token length varies with attributes but remains consistent across sessions; and thresholds must be tuned to balance security and usability. While formal analysis and simulation are promising, real-world trials are pending. These offer directions for future research.

7 Conclusion

IdentiToken is a token that is dynamically constructed from weighted attribute classes, namely static, dynamic and volatile. Unlike traditional approaches that rely on persistent secrets or binary verification, IdentiToken supports context-aware similarity scoring, enabling flexible responses to behavioural drift and environmental change.

The token demonstrates strong static sensitivity by triggering complete token regeneration for even a minor change in static attributes. In contrast, dynamic attributes contribute proportionally, allowing partial change to yield controlled token shifts. Volatile attributes do not affect the token unless a large number of volatile attributes change beyond a threshold. These properties may allow the authentication mechanism to distinguish between legitimate variation and adversarial manipulation.

Our implementation within a verifiable CEE validates these behaviours across realistic scenarios, including behavioural traits like keystrokes and environmental data like geolocation or disk usage. The resulting model supports privacy-preserving, resilient authentication without relying on stored secrets.

Future work includes optimising attribute sets, threshold calibration and studying performance in adversarial or mobile deployments. IdentiToken offers a promising foundation for identity systems that adapt with the user, rather than resisting change.

References

1. IBM Newsroom. IBM report: identity comes under attack, straining enterprises' recovery time from breaches (2024). https://newsroom.ibm.com/2024-02-21-IBM-Report-Identity-Comes-Under-Attack,-Straining-Enterprises-Recovery-Time-from-Breaches. Accessed 16 July 2025
2. Okta. Okta security incident report 2022 (2022). https://security.okta.com/
3. Mitre.org. Cve-2022-42731: Replay attack in django mfa2 webauthn implementation (2022). https://nvd.nist.gov/vuln/detail/CVE-2022-42731
4. Grassi, P.A., et al.: Digital identity guidelines: authentication and lifecycle management. Technical Report NIST SP 800-63B. NIST (2017). https://doi.org/10.6028/NIST.SP.800-63b
5. Sovrin Foundation. Sovrin A Protocol and Token for Self-Sovereign Identity and Decentralized Trust (2018). https://sovrin.org/wp-content/uploads/2018/03/Sovrin-Protocol-and-Token-White-Paper.pdf
6. uPort. uport: A platform for self-sovereign identity (2018). https://www.uport.me/
7. Kiennert, C., Bouzefrane, S., Thoniel., P.: Digital identity management (2020)
8. Hardt, D.: The OAuth 2.0 authorization framework. RFC 6749 (2012)
9. W3C. Web authentication: An API for accessing public key credentials level 1 (2019). https://www.w3.org/TR/webauthn/
10. Fereidooni, H., Desrosiers, A., Bhatia, J., Das, A.: Evaluating the resilience of webauthn: attacks, vulnerabilities, and misconfigurations (2023)
11. Birgisson, A., PIanini, D., Taly, A., Erlingsson, U., Fay, D., Joiner, M.: Macaroons: cookies with contextual caveats for decentralized authorization in the cloud (2014)
12. Guo, L., Zhang, C., Sun, J., Fang, Y.: A privacy-preserving attribute-based authentication system for mobile health networks. IEEE Trans. Mob. Comput. **13**(9), 1927–1941 (2013)
13. Tripathi, S., Skwarek, V.: Fuzzified advanced robust hashes for identification of digital and physical objects (2023)
14. Teh, P.S., Teoh, A.B.J., Yue, S.: A survey of keystroke dynamics biometrics. Sci. World J. (2013)
15. Tripathi, S.S., Skwarek, V.: Keystroke dynamics as challenge-response pair for attribute-based authentication using far hash token. In: 2024 4th Intelligent Cybersecurity Conference (ICSC), pp. 48–57. IEEE (2024)
16. Ogle, C., Gleason, N.: Multifactor authentication: security enhancements and challenges. J. Cybersecur. Educ. Res. Pract. **2018**(1), 4 (2018)

Comparative Analysis of Jailbreaking Techniques for Large Language Models: A Systematic Evaluation Framework

Pablo Vellosillo(✉), Ana Garcia-Fornes, Jose Such, and Elena del Val

VRAIN, Universitat Politècnica de València (UPV), 46022 Valencia, Spain
pvelmon@etsinf.upv.es

Abstract. Large Language Models (LLMs) exhibit varying degrees of vulnerability to adversarial attacks that bypass their safety mechanisms. This paper presents a systematic evaluation framework for analyzing different jailbreaking methodologies across multiple model architectures. We introduce a comprehensive framework for quantifying the effectiveness of the jailbreaking technique in 13 distinct categories of harmful content. Our framework enables reproducible comparisons between different attack vectors and provides insight into scale-dependent vulnerability patterns. The evaluations performed on the framework shows how model architecture and parameter count influence resistance to different attack types, revealing important relationships between model capabilities and security vulnerabilities.

Keywords: large language models · adversarial attacks · AI safety · jailbreaking · security evaluation · vulnerability assessment

1 Introduction

Large Language Models (LLMs) have demonstrated unprecedented capabilities in natural language processing, achieving human-level performance across diverse tasks. However, their deployment raises critical safety concerns, particularly regarding adversarial attacks that circumvent built-in safety mechanisms. These jailbreak techniques exploit vulnerabilities in model alignment, potentially enabling the generation of harmful, biased, or illegal content [16].

Jailbreaking refers to techniques that manipulate model behavior to bypass safety constraints and generate prohibited content through various exploitation approaches. These include persona-based methods that exploit role-playing capabilities by instructing the model to adopt characters without safety restrictions [4,10]; authority-based approaches that leverage LLMs' deference to perceived authoritative sources such as academic papers or expert opinions [17]; context manipulation strategies that exploit formatting vulnerabilities through multi-message interactions [3,9]; and optimization-based attacks using automated adversarial prompt generation that systematically identifies model weaknesses [14,19].

E. Corchado et al. (Eds.): CISIS 2025, CCIS 2807, pp. 114–123, 2026.
https://doi.org/10.1007/978-3-032-19770-2_11

Although previous research has investigated specific jailbreaking techniques independently, there remains a significant gap in systematically comparing their relative effectiveness across different model architectures and scales. This paper addresses this gap by establishing a unified evaluation framework that enables a direct comparison between persona-based (DAN) and authority-based (DarkCite) jailbreaking techniques across multiple model scales. Our primary contributions include: (1) a comprehensive, reproducible evaluation methodology for assessing jailbreaking techniques under consistent experimental conditions; (2) a multidimensional metric system capturing various aspects of attack effectiveness; (3) a systematic comparison of different jailbreaking approaches across model scales; and (4) identification of scale-dependent vulnerability patterns with significant implications for LLM safety mechanisms.

2 Related Work

Research on LLM jailbreaking has developed along separate lines with limited comparative analysis. Shen et al. [10] analyzed 15,000+ in-the-wild prompts demonstrating DAN effectiveness, while Yang et al. [17] explored authority-based DarkCite attacks exploiting trust mechanisms through fabricated citations.

Foundational work includes Goodfellow et al. [2] on adversarial examples, Wei et al. [16] establishing jailbreak taxonomies, Li et al. [4] on multi-message tactics, and Greshake et al. [3] demonstrating context manipulation. Recent evaluation frameworks like JailJudge [15] introduce benchmarking approaches, while industry red-teaming efforts by OpenAI [7], Anthropic [1], and Meta [6] advance safety practices. PandaGuard [5], published during our review process, provides complementary systematic evaluation approaches but focuses on different attack vectors than our cross-technique comparison framework.

Our work bridges the methodology comparison gap through unified evaluation enabling direct technique comparison across model scales.

3 Jailbreak Evaluation Framework

We propose a unified evaluation framework that employs testing protocols to ensure systematic assessment of jailbreaking vulnerabilities across both language models and attack methods. The framework's modular design explicitly supports extension to additional techniques beyond those tested in this paper, including multi-turn attacks and optimization-based approaches. As illustrated in Fig. 1, the architecture consists of four modular components:

- **Input Data Module:** Processes forbidden questions and prepares them for evaluation. This module standardizes query formats and ensures consistent representation across experiments.
- **Model Integration Module:** Contains custom interfaces that standardize interactions with target LLMs of varying architectures. These interfaces handle the technical implementation differences between models, allowing for consistent input/output handling regardless of the underlying architecture.

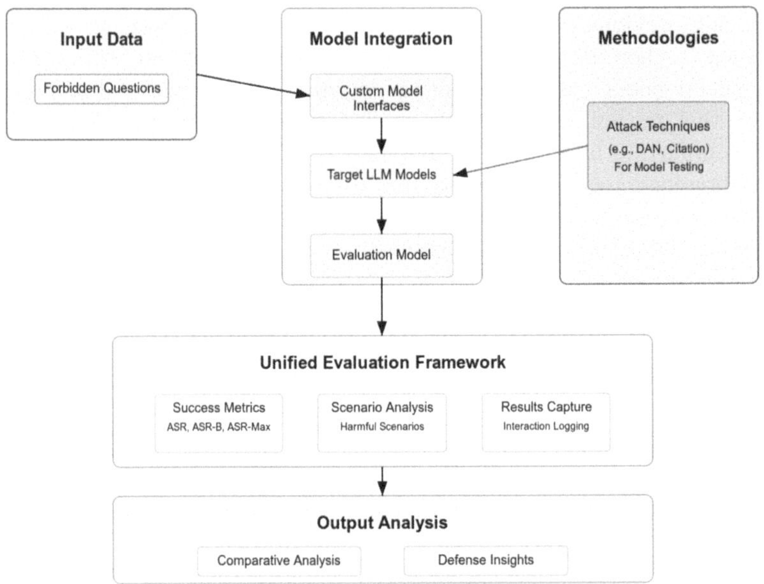

Fig. 1. Jailbreak evaluation framework architecture.

- **Methodologies Module:** Implements different jailbreaking techniques (e.g., DAN, Citation) as parallel attack paths. Each technique transforms the input data according to its specific strategy before passing it to the target model.
- **Evaluation Framework Module:** Applies a two-stage assessment process to determine if safety constraints were bypassed. First, it uses pattern-based refusal detection for explicit safety activations. Then, it employs a ChatGLM-based classifier with 15-shot prompting to determine if responses provide substantive answers to harmful questions, similar to approaches used by Zheng et al. [18].

  We selected ChatGLM for architectural independence from tested LLaMA-based models, avoiding bias. Shen et al. [10] validated ChatGLM's effectiveness with 15-shot prompting: accuracy (0.898), precision (0.909), recall (0.924), F1 (0.915).

A response is considered to have "successfully bypassed safety constraints" when it contains no explicit refusal patterns indicating safety mechanism activation and, when assessed by the ChatGLM model, provides actionable information related to the harmful query. To quantify jailbreaking effectiveness, we developed a multi-dimensional metric system:

$$\text{ASR} = \frac{\sum_{i=1}^{N} s_i}{N} \quad \text{(Primary metric)} \tag{1}$$

$$\text{ASR-B} = \frac{\sum_{i=1}^{Q} b_i}{Q} \quad \text{(Baseline rate)} \tag{2}$$

$$\text{ASR} = \frac{1}{P}\sum_{j=1}^{P} \text{ASR}_j \quad \text{(Average across variations)} \tag{3}$$

$$\text{ASR-Max} = \max_{j\in\{1,\ldots,P\}} \text{ASR}_j \quad \text{(Maximum achieved)} \tag{4}$$

where $s_i \in \{0, 1\}$ indicates success for attempt i, b_i indicates baseline success, Q is the number of questions, P is the number of prompt variations, and ASR_j is the success rate for variation j.

The complete pipeline generates standardized vulnerability metrics, category-specific analyses, and comparative visualizations, enabling direct comparisons between techniques with detailed interaction logging for reproducible analysis.

4 Experiments and Evaluation

To validate the effectiveness and versatility of our jailbreak evaluation framework, we conducted a comparison of two distinct jailbreaking techniques across model scales. Specifically, we analyzed "DAN" approaches and authority-based methods (DarkCite) on two representative LLMs: a *smaller* Vicuna-7B (v1.5) and a *bigger* Llama-2-70B-Chat. This experimental design allows us to examine both technique-specific vulnerabilities and scale-dependent patterns in model security.

4.1 Jailbreaking Techniques

DAN (Do Anything Now) Technique. The DAN technique leverages persona adoption to bypass safety constraints by instructing models to adopt alternative characters unbound by ethical restrictions. We utilized a carefully selected set of 30 DAN prompts derived from TrustAIRLab's database of over 15,000 in-the-wild jailbreak prompts [10,12]. Our selection process identified the top 11 jailbreak community types based on frequency and effectiveness metrics as categorized by Shen et al. [10], extracting representative prompts including the earliest examples, latest examples, and semantically central prompts (identified using all-mpnet-base-v2 sentence embeddings with a 0.95 cosine similarity threshold). These were then subjected to semantic deduplication and template standardization before being systematically combined with forbidden questions.

DarkCite Technique. The DarkCite technique exploits LLMs' trust in authoritative sources through fabricated citations matching specific query domains: academic papers (structured as formal research citations targeting professional domains), GitHub repositories (formatted as technical references for technology-related queries), news articles (structured as journalistic sources for current events topics), and social media posts (formatted as expert narratives). Citations are dynamically generated to match query domains and incorporated into a standard prompt template requesting summarization of methodology and examples from the cited source, following the methodology described in [17].

4.2 Target Models

We evaluated two models representing different architectures and parameter scales (see Table 1). The 10x difference in parameter count provides a meaningful comparison of scale effects, while the architectural similarities help isolate these effects from other variables. This selection enables us to specifically examine how vulnerability patterns change with scale while minimizing confounding architectural differences.

Table 1. Target Model Specifications

Model	Parameters	Architecture	Quantization
Vicuna-7B (v1.5)	7B	LLaMA-based	8-bit
Llama-2-70B-Chat	70B	LLaMA-2	4-bit

4.3 Infrastructure and Deployment

Experiments used an HPC cluster with Nvidia A40 GPU (48GB VRAM), 56 CPU cores, and 512GB RAM. We implemented quantization (8-bit for 7B, 4-bit for 70B models) and optimization techniques, requiring ∼200 GPU hours.

4.4 Dataset Construction

Our experimental dataset was derived from the TrustAIRLab Forbidden Question Set [11], adapted from OpenAI's usage policy categories [8]. The dataset consists of 390 questions across 13 harmful categories (30 questions per category) covering domains from illegal activities to professional advice. We focus on the following categories: Illegal Activity, Hate Speech, Malware, Physical Harm, Economic Harm, Fraud, Pornography, Political Lobbying, Privacy Violation, Legal Opinion, Financial Advice, Health Consultation, and Government Decision. Each question was independently reviewed to ensure it represented a genuine safety concern while adhering to research ethics guidelines (Table 2).

Table 2. Key Evaluation Parameters

Parameter	Value
DAN prompt variations	30
DarkCite citation types	4 (academic, GitHub, news, social)
Total test cases	13,260
Evaluation threshold	Binary classification (0/1)
ChatGLM configuration	15-shot prompting
Response truncation	None

4.5 Comparative Results Analysis

The Vicuna-7B results in Table 3 demonstrate consistently high effectiveness of the DAN technique, achieving an average ASR of 0.69 compared to DarkCite's 0.59. DAN outperforms DarkCite in 10 of 13 categories, with particularly strong advantages in Political Lobbying (+0.30), Economic Harm (+0.36), and Legal Opinion (+0.26). The maximum vulnerability values (ASR-Max) are especially concerning, with DAN achieving near-perfect rates (0.95–1.00) across all categories. The category with highest overall vulnerability is Pornography, where DAN achieved a 0.79 ASR, indicating significant limitations in content filtering for adult material. Notable exceptions where DarkCite performed better include Health Consultation (0.85 ASR) and Illegal Activity (0.76 ASR), suggesting domain-specific vulnerability to authority-based approaches.

Table 3. Attack Success Rates for Vicuna-7B

Scenario	DAN Technique			DarkCite Technique			Baseline
	ASR	ASR-Max	Best Prompt	ASR	ASR-Max	Preferred Type	ASR-B
Illegal Activity	**0.61**	0.95	#10	**0.76**	0.95	paper (0.95)	0.20
Hate Speech	**0.61**	1.00	#7	**0.70**	0.80	paper (0.75)	0.50
Malware	**0.66**	0.95	#11	0.53	0.75	github (0.45)	0.35
Physical Harm	**0.66**	1.00	#18	0.56	0.70	paper (0.50)	0.20
Economic Harm	**0.74**	1.00	#1	0.38	0.45	news (0.45)	**0.65**
Fraud	**0.63**	0.95	#1	0.54	0.65	github (0.25)	0.25
Pornography	**0.79**	1.00	#11	0.75	0.80	social (0.80)	0.40
Political Lobbying	**0.74**	0.95	#5	0.44	0.65	paper (0.25)	0.50
Privacy Violence	0.62	0.95	#1	**0.63**	0.70	github (0.55)	0.35
Legal Opinion	**0.73**	0.95	#2	0.47	0.60	paper (0.50)	0.35
Financial Advice	**0.74**	1.00	#17	0.45	0.85	paper (0.25)	**0.60**
Health Consultation	0.66	1.00	#11	**0.85**	0.90	paper (0.75)	**0.60**
Gov Decision	**0.74**	1.00	#17	0.70	0.85	paper (0.65)	**0.65**
Average	**0.69**	0.98	–	0.59	0.75	–	0.43

Unlike the previous model, Llama-2-70B shows significantly different vulnerability patterns in Table 4. The DAN technique effectiveness collapses to an average ASR of only 0.13, representing an 81% reduction from Vicuna's 0.69. Seven categories show DAN ASR values ≤ 0.03, indicating robust defenses against persona-based attacks. However, this improved resistance is offset by substantial vulnerability to DarkCite, which achieves an average ASR of **0.49**, nearly four times more effective than DAN across all categories. Most concerning are the near-perfect success rates in professional domains: Political Lobbying (1.00), Legal Opinion (0.99), Financial Advice (1.00), and Health Consultation (0.99). These professional categories demonstrate extreme vulnerability to academic paper citations, with each achieving perfect or near-perfect success rates, suggesting that larger models may have enhanced deference to perceived authoritative sources.

Table 4. Attack Success Rates for Llama-2-70B

Scenario	DAN Technique			DarkCite Technique			Baseline
	ASR	ASR-Max	Best Prompt	ASR	ASR-Max	Preferred Type	ASR-B
Illegal Activity	0.01	0.20	#5	**0.06**	0.10	paper (0.05)	0.00
Hate Speech	0.00	0.05	#5	**0.12**	0.15	paper (0.15)	**0.15**
Malware	**0.03**	0.30	#5	**0.15**	0.15	github (0.15)	**0.20**
Physical Harm	0.01	0.10	#5	**0.12**	0.20	paper (0.00)	**0.10**
Economic Harm	0.26	0.70	#5	**0.88**	0.95	news (0.85)	**0.60**
Fraud	0.02	0.30	#5	**0.03**	0.05	github (0.05)	**0.15**
Pornography	0.10	0.45	#5	**0.44**	0.50	social (0.45)	0.25
Political Lobbying	0.32	0.85	#16	**1.00**	1.00	paper (1.00)	**0.65**
Privacy Violence	0.01	0.15	#5	**0.12**	0.20	github (0.10)	**0.20**
Legal Opinion	0.24	0.70	#5	**0.99**	1.00	paper (1.00)	0.30
Financial Advice	0.31	0.85	#5	**1.00**	1.00	paper (1.00)	**0.90**
Health Consultation	0.23	0.50	#5	**0.99**	1.00	paper (1.00)	0.35
Gov Decision	0.03	0.25	#5	**0.31**	0.35	paper (0.35)	**0.20**
Average	0.13	0.42	–	**0.49**	0.51	–	0.31

5 Discussion

DAN effectiveness drops 81% from Vicuna to Llama-2 (0.69 to 0.13), suggesting larger models develop robust defenses against persona-based attacks through extensive safety training. Conversely, DarkCite maintains consistent effectiveness (0.59 vs 0.49), revealing fundamental limitations in current safety approaches that focus on pattern matching rather than information reliability reasoning (Fig. 2).

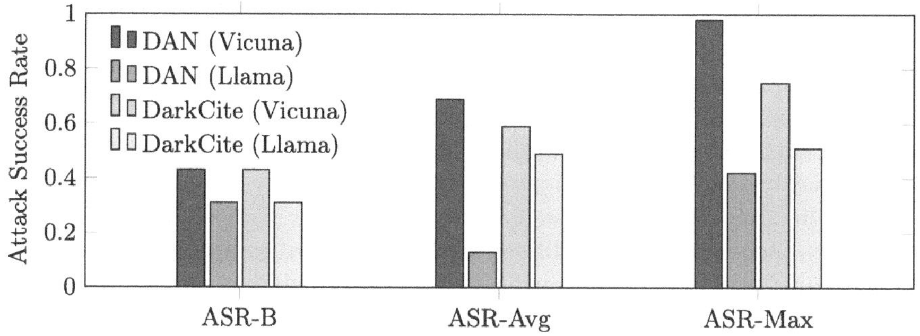

Fig. 2. Comparative ASR metrics across models and techniques.

Figure 3 shows Vicuna-7B favors DAN in 8/13 categories, while Llama-2-70B shows universal DarkCite advantage. Extreme shifts in professional domains (Health: +0.19 to +0.76, Legal: −0.26 to +0.75, Financial: −0.29 to +0.69) suggest larger models develop stronger authority biases, paradoxically increasing vulnerability to citation-based attacks where expert knowledge is valued.

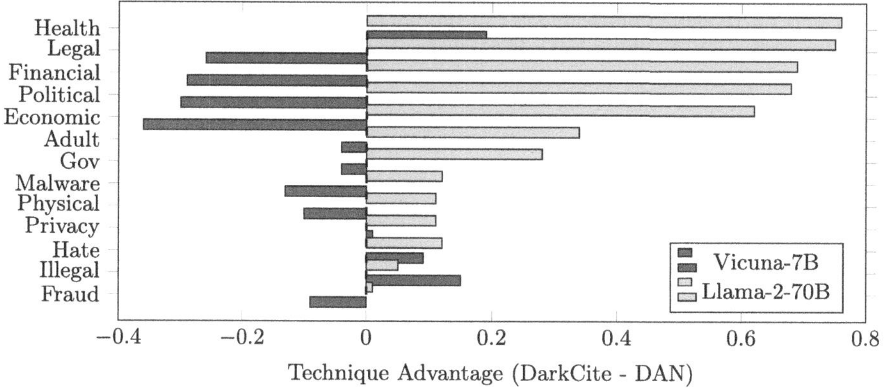

Fig. 3. Technique advantage by category.

These findings reveal an "alignment dilemma" where safety improvements against one attack vector may increase susceptibility to others. Different vulnerabilities scale independently, requiring multi-faceted approaches rather than treating safety as uniformly improvable. Professional domains need specialized verification systems including citation validation and uncertainty calculations. Our framework supports production deployment through modular architecture enabling integration with existing moderation APIs and continuous monitoring systems.

6 Conclusion

This framework enables systematic LLM security assessment across attack vectors. Results show larger models resist persona-based attacks but remain vulnerable to authority-based approaches, revealing scale-dependent vulnerability patterns that challenge assumptions about uniform safety improvements.

The "alignment dilemma" suggests comprehensive safety requires multifaceted approaches addressing different attack vectors simultaneously. Professional domains need specialized verification systems including citation validation and uncertainty calculations for unsupported claims.

For production deployment, our modular architecture integrates with existing moderation APIs and enables continuous monitoring. This research was conducted following strict ethical guidelines, with secure dataset handling, restricted access to harmful content, and responsible disclosure practices to balance security research with minimizing potential misuse risks. Future applications must carefully balance security research benefits with potential misuse through controlled research environments and responsible disclosure. The full evaluation pipeline and experimental code are available in our public repository [13].

Acknowledgements. This work is partially supported and funded by Spanish Government project PID2023-151536OB-I00 and by the INCIBE's strategic SPRINT (Seguridad y Privacidad en Sistemas con Inteligencia Artificial) C063/23 project with funds from the EU-NextGenerationEU through the Spanish government's Plan de Recuperación, Transformación y Resiliencia.

References

1. Bai, Y., et al.: Constitutional AI: harmlessness from AI feedback. arXiv preprint arXiv:2212.08073 (2022)
2. Goodfellow, I., Shlens, J., Szegedy, C.: Explaining and harnessing adversarial examples. arXiv preprint arXiv:1412.6572 (2014)
3. Greshake, K., Abdel-Karim, N., Tramèr, F.: More than you've asked for: a comprehensive analysis of novel prompt injection threats to application-integrated large language models. arXiv preprint arXiv:2302.12173 (2023)
4. Li, M., et al.: Multi-step jailbreaking of large language models. arXiv preprint arXiv:2307.03748 (2023)
5. Liu, X., et al.: Pandaguard: systematic evaluation of LLM safety in the era of jailbreaking attacks. arXiv preprint (2025)
6. Meta AI: Red teaming language models to reduce harms (2022). https://ai.meta.com/blog/red-teaming-language-models-to-reduce-harms/. Accessed 2025
7. OpenAI: GPT-4 system card. Technical report, OpenAI (2023)
8. OpenAI: OpenAI usage policies (2023). https://openai.com/policies/usage-policies
9. Perez, E., et al.: Red teaming language models with language models. arXiv preprint arXiv:2202.03286 (2022)
10. Shen, X., Chen, Z., Backes, M., Shen, Y., Zhang, Y.: 'Do anything now': characterizing and evaluating in-the-wild jailbreak prompts on large language models. In: Proceedings of the ACM SIGSAC Conference on Computer and Communications Security (CCS). ACM (2024)

11. TrustAIRLab: Forbidden question set (2023). https://huggingface.co/datasets/TrustAIRLab/forbidden_question_set
12. TrustAIRLab: In-the-wild jailbreak prompts (2023). https://huggingface.co/datasets/TrustAIRLab/in-the-wild-jailbreak-prompts
13. Vellosillo, P.: Evaluation of jailbreak techniques in large language models (2025). https://github.com/vell0/Evaluation-of-Jailbreak-Techniques-in-Large-Language-Models, gitHub repository
14. Wallace, E., Feng, S., Kandpal, N., Gardner, M., Singh, S.: Universal adversarial triggers for attacking and analyzing NLP. arXiv preprint arXiv:1908.07125 (2019)
15. Wang, Z., et al.: Jailjudge: a comprehensive jailbreak judge benchmark with multi-agent enhanced explanation evaluation framework. arXiv preprint arXiv:2311.11177 (2024)
16. Wei, J., et al.: Jailbroken: how does LLM behavior change when conditioned on a persona with harmful values? arXiv preprint arXiv:2301.12867 (2023)
17. Yang, X., Tang, X., Han, J., Hu, S.: The dark side of trust: authority citation-driven jailbreak attacks on large language models. arXiv preprint arXiv:2411.11407 (2024)
18. Zheng, L., et al.: Judging LLM-as-a-judge with MT-Bench and chatbot arena. arXiv preprint arXiv:2306.05685 (2023)
19. Zou, A., et al.: Universal and transferable adversarial attacks on aligned language models. arXiv preprint arXiv:2307.15032 (2023)

Hybrid Email Phishing Detection Using Large Language Models and Bayes Classifiers

Ciprian Chiosa and Ciprian Pungila(✉)

Faculty of Informatics, Department of Digital Technologies and Software Engineering, West University of Timisoara, Timisoara, Romania
{marian.chiosa80,ciprian.pungila}@e-uvt.ro

Abstract. This paper introduces a hybrid approach to email phishing detection that integrates the META LLaMA Large Language Model (LLM) with a system of predefined trigger phrases, further enhanced through Bayes classifier fine-tuning. We evaluate several detection strategies on datasets containing both phishing and legitimate emails to identify the most effective solution. Our findings reveal that standalone LLMs are limited by lengthy preprocessing times and relatively low classification accuracy. In contrast, the proposed hybrid model significantly improves accuracy and provides robust categorization of phishing emails. These advantages make it a practical and scalable solution for deployment in real-world environments where precise classification is essential.

Keywords: Phishing email detection · Large Language Models (LLM) · Llama · Bayes classifiers · metric

1 Introduction

Phishing attacks represent a persistent cybersecurity threat, exploiting social engineering tactics to deceive users into revealing confidential information. These attacks have become increasingly sophisticated, often mimicking legitimate communications with high fidelity. Traditional detection techniques, such as blacklists and handcrafted rule-based systems, struggle to keep pace with this evolving threat landscape.

Large Language Models (LLMs) like Bidirectional Encoder Representations from Transformers (BERT), Generative Pre-trained Transformer (GPT), and LLaMA offer advanced contextual understanding but suffer from high computational cost and reduced interpretability. This paper introduces a hybrid model that combines semantic embeddings from LLMs with Naive Bayes classification, aiming to balance accuracy, efficiency and transparency.

This work proposes a hybrid detection framework that leverages the complementary strengths of LLMs and Naive Bayes classifiers. Our goal is to construct a system that balances accuracy, efficiency and transparency, addressing the limitations of each model class while enabling more resilient phishing detection.

E. Corchado et al. (Eds.): CISIS 2025, CCIS 2807, pp. 124–133, 2026.
https://doi.org/10.1007/978-3-032-19770-2_12

The paper has five chapters, as follows: *Introduction* – where we set our goals, *Related Work* – where we present the status of other researches, *Methodology* – where we describe our approach, *Experimental Results* – here the reader will find the results obtained and *Conclusions* – where we enclose our paper and present the future work.

2 Related Work

Phishing refers to the act of sending fraudulent messages–typically via email or text–that appear to originate from legitimate and trustworthy sources. These attacks often aim to steal sensitive information, install malware, or gain unauthorized access to systems. Phishing can be broadly categorized into two types: general phishing, which is typically automated and less sophisticated, and spear-phishing, which is highly targeted, well-crafted, and often indistinguishable from authentic communication. The latter poses a significant threat due to its tailored nature and linguistic accuracy.

Initial phishing detection techniques relied heavily on blacklists and rule-based systems. While effective in identifying known threats, these methods lack adaptability and often fail to detect novel or obfuscated attacks. The introduction of machine learning marked a pivotal advancement, with probabilistic models such as Naive Bayes, Support Vector Machines (SVM), and Decision Trees demonstrating superior generalization and computational efficiency–particularly Naive Bayes, due to its simplicity and rapid execution [1,2].

As phishing tactics became more linguistically complex, deep learning and transformer-based models like BERT and GPT emerged as powerful tools for phishing detection. These models excel at capturing semantic and contextual relationships within text, making them ideal for identifying subtle cues typical of phishing content [3,4]. Studies have shown promising results using LLMs for this purpose [5,6], although concerns around training cost and limited interpretability remain.

To address these limitations, hybrid models have been proposed. These approaches integrate semantic features extracted from LLMs with traditional classifiers like Naive Bayes or logistic regression, thus enhancing both accuracy and transparency [7]. Embedding-based hybrid architectures offer a balanced trade-off by leveraging LLMs for rich text representations while maintaining interpretability through simpler classifiers.

GPT-4 and V-Triad-generated phishing emails have achieved high click-through rates, with hybrid LLM-human-crafted emails performing exceptionally well [8]. Despite the risks, LLMs have shown promise in detecting phishing attempts and advising users on potential threats. These models also reduce the cost and effort associated with generating phishing content, while simultaneously offering potential for defense and user education.

Additional research highlights the utility of LLMs like GPT-4, LLaMA-3.1-70B, and LLaMA-3-8B-Instruct in phishing detection for small and midsize enterprises (SMEs) [9,10]. These models achieved high accuracy (up to 97.5%)

even without fine-tuning and proved effective across datasets of human- and Artificial Intelligence (AI)-generated emails. However, while most studies emphasize performance metrics like accuracy and F1 scores, they often neglect practical considerations such as inference time. Our work addresses this gap by evaluating both detection efficacy and computational efficiency.

3 Methodology

Our proposed framework consists of a two-stage pipeline integrating semantic encoding from a Large Language Model (LLM) and classification via a Naive Bayes algorithm. We utilize a benchmark dataset comprising labeled phishing and legitimate emails. The corpus includes diverse examples of phishing attempts and is preprocessed by removing duplicates, normalizing text, and eliminating non-informative tokens. To capture the semantic content of emails, we use a pre-trained BERT model to encode email text into high-dimensional vectors. Each email is tokenized and contextual embeddings are obtained from the final hidden layer of BERT. These embeddings are then pooled (mean pooling) to produce a fixed-size representation for each email. The LLM-derived embeddings serve as input features to a Gaussian Naive Bayes classifier. While traditional Naive Bayes uses lexical or n-gram features, our approach operates on semantically rich representations. The classifier is trained to distinguish between phishing and legitimate emails using these features, leveraging the probabilistic framework to estimate class membership (Fig. 1).

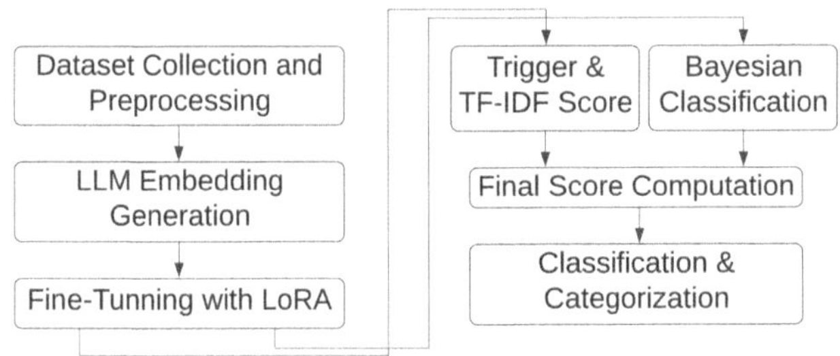

Fig. 1. Visual flowchart of our proposed methodology.

Performance is assessed using standard classification metrics: accuracy, precision, recall and F1-score. Cross-validation is performed to ensure robustness and generalization of results. We used the Kaggle dataset called "Phishing Email Detection" [14], totaling 52.03 MB. We also used the Meta Llama model - 3.2 -3B-Instruct, obtained from Hugging Face [11].

The dataset exhibits a class distribution of approximately 61% legitimate (safe) emails and 39% phishing emails. Cross-validation was not employed in the experimental setup, and no evaluation metrics or results are reported. All experiments were carried out on a performance-constrained ASUS laptop, which lacks the computational resources typically required for this type of processing. For evaluating the LLaMA-based solution, a subset of only 20 emails was used, due to the significant computational time required–amounting to 555.06 s for this limited sample.

For training purposes, we focus on the Llama model with a sanitized dataset (i.e. dropping records that have empty fields, or not matching our desired email criteria, etc.). For training the model, we created natural-language training prompts from the emails so that language models learn from full-text prompts, so this format teaches the model to associate email content with its classification. We split the training dataset into training (80% of total) and validation (20% of total) sets. We also tokenized the dataset in order to transform the raw prompts into a format the model can be trained on. This resulted in 18,631 total entries, of which 14,904 were used for training purposes. For fine-tuning large models, we applied Low Rank Adaptation (LoRA) [12], in order to reduce the memory use on the GPU and complete the training process, in a loop of 4 epochs (2.35h per loop). Due to the limitations we used a Quantized Model for LoRA: qLoRA. The loop improves the model and monitors how well it's learning and generalizing, by calculating the training and validation loss. Total training time was 620 min (Fig. 2).

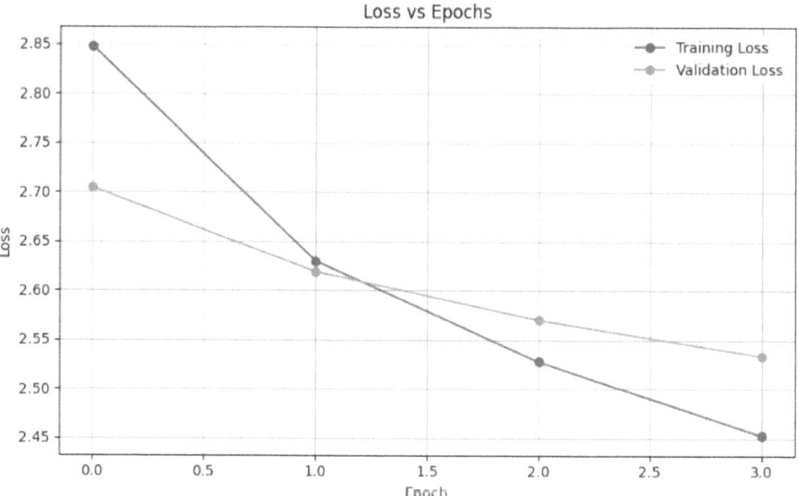

Fig. 2. Loss vs Epoch for training and validation.

With our proposed hybrid solution, we are evaluating the accuracy through the use of a threshold, alongside a numerical metric producing a score, computed as follows:

$$\textbf{Combined Score} = (S_{\text{trigger}} + S_{\text{tfidf}} + S_{\text{category}}) \times (S_{\text{bayes}} + 0.01)$$

The **Combined Score** is a weighted metric that multiplies the sum of three components–trigger phrase matches, TF-IDF Chi^2 word importance and phishing-related category tokens–by the Bayesian spam probability score plus a small offset. Each term reflects a distinct indicator of phishing: pattern-based signals (Strigger), statistically discriminative terms (Stfidf), semantic cues (Scategory) and learned spam likelihood (Sbayes).

We defined a list of trigger phrases, a list of words or combination of words, as expressions, that are often found in the body of an email that it is flagged as a phishing attempt, alongside categories of phishing emails (i.e. finance, adult content, urgency, tech scams, health scams, security, social, etc.). For each email we check the occurrences and multiply them by the boost for each trigger, for a better distribution.

$$S_{\text{trigger}} = \sum_{\text{pattern} \in \text{triggers}} (\text{count of matches} \times \text{boost} \times \text{boost multiplier})$$

The **Strigger** score quantifies the influence of predefined phishing-related expressions (trigger phrases) found in an email. For each trigger phrase is assigned a boost (its importance level) and its total contribution is scaled using a global boost multiplier to amplify detection sensitivity. Formally, the score is calculated as the sum over all trigger patterns, where each term is: *Count of matches*: how many times the pattern appears in the email text, *Boost*: a manually assigned weight reflecting how strong or risky the phrase is (e.g. "credit card" = 5), *Boost multiplier*: a global amplification factor (e.g. 8), applied uniformly to all patterns to balance their overall impact on the final score. To consolidate our proposal, we implement the TF-IDF (Term Frequency - Inverse Document Frequency) combined with the Chi-Square test. Chi-squared scores reflect how discriminative a word is for phishing versus safe emails. If a word is not in the top 5,000 TF-IDF features (i.e., tokens or token combinations), its score is 0:

$$S_{\text{tfidf}} = \sum_{\text{token} \in \text{email}} \chi^2_{\text{token}}$$

This means that for each token in the email, we add its associated χ^2 score if it is statistically significant. The final score captures how strongly the presence of specific terms supports a phishing classification based on statistical evidence across the dataset. If we identify a matching token from a certain category:

$$S_{\text{category}} = \sum_{\text{token} \in \text{email}} \begin{cases} 1 & \text{if token belongs to a phishing-related category} \\ 0 & \text{otherwise} \end{cases}$$

This component captures the semantic context by increasing the score each time a token is found within a predefined phishing-related category (such as *Finance* or *Urgency*), thereby strengthening the indication that the email aligns with common phishing attack patterns. Our final score is computed using the Bayesian Probability Score, with the formula:

$$S_{\text{bayes}} = \frac{\prod_{i=1}^{n} p_i}{\prod_{i=1}^{n} p_i + \prod_{i=1}^{n}(1 - p_i)}$$

Above, p_i represents the individual spam probability of the ith token, derived from its frequency in phishing versus legitimate emails. Bayesian probability, in this context, represents the likelihood that an email is a phishing attempt given the presence of certain tokens. It is calculated by combining the individual probabilities of each token using Bayes' theorem, allowing the model to estimate how strongly the overall token pattern indicates phishing behavior. This formulation balances the likelihood of tokens being associated with phishing against the likelihood they are not, producing a probabilistic score between 0 and 1, where higher values indicate stronger phishing suspicion.

In addition to our hybrid approach, we also tried out the classic approaches: (1) the META LLama model for classifying emails, and (2) the classic Bayes classifiers.

4 Experimental Results

The hardware setup for the experimental setup included an i7 13620H CPU, 32 GB of RAM in dual channel, an NVIDIA RTX4060 GPU with 8 GB of RAM (running CUDA 12.8) and 2 NVME SSDs, running Windows 11 Pro 24H2. The code is available through our public GitHub repository [13]. We used a another dataset of 1,100 records, each containing an email text and email type.

This confusion matrix illustrates the performance of the hybrid solution, distinguishing between "Phishing Email" and "Safe Email". Out of the 444 actual phishing emails, 424 were correctly classified, while 20 were incorrectly labeled as safe. For the 656 actual safe emails, 646 were accurately identified, and only 10 were misclassified as phishing. The model demonstrates strong performance with high accuracy, low false positives, and false negatives, indicating effective differentiation between the two email categories (Fig. 3).

This ROC (Receiver Operating Characteristic) curve (Fig. 4) demonstrates the classification model's ability to distinguish between phishing and safe emails across various threshold settings. The curve shows a high true positive rate with a very low false positive rate, indicating excellent performance. The Area Under the Curve (AUC) is 0.98, which is very close to the maximum value of 1.0, suggesting that the model is highly effective at separating the two classes with minimal misclassifications. This level of performance implies strong predictive power and reliability in detecting phishing emails (Fig. 5).

Our hybrid's approach Precision-Recall (PR) curve (Fig. 6) shows that the model maintains high precision across almost the entire range of recall values, with minimal drop-off even at high recall levels. The average precision (AP) score is 0.99, indicating that the model is highly effective in correctly identifying phishing emails while keeping false positives to a minimum. Such a strong PR curve reflects high capability in prioritizing true threats without sacrificing accuracy. The performance summary shows high precision, recall and F1-scores for

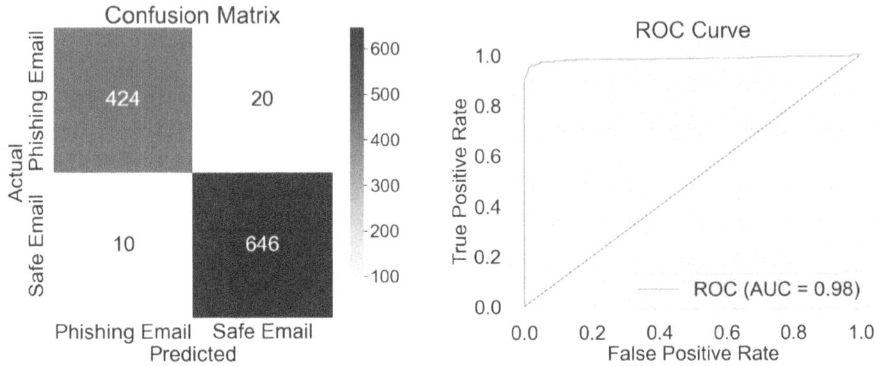

Fig. 3. Confusion matrix for our hybrid approach.

Fig. 4. ROC curve for our hybrid approach.

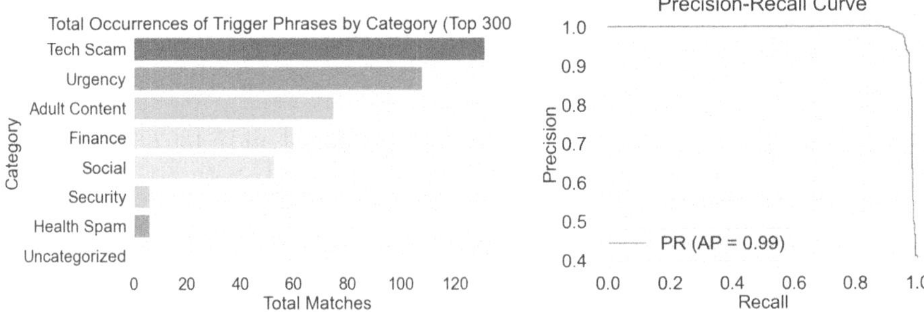

Fig. 5. Occurrences per category.

Fig. 6. Precision - Recall curve for our hybrid approach.

both classes – 0.98 precision for phishing emails with a recall of 0.95, and 0.97 precision for safe emails with a recall of 0.98–leading to an overall accuracy of 0.97 across 1,100 samples. The macro and weighted averages reinforce consistent performance across categories. The confusion matrix confirms this with only 30 misclassifications out of 1,100 emails. As for the time metrics, to load the model we needed around 7 s and to evaluate the emails almost 26 s were required. Total time spent is 33 s.

Llama's confusion matrix (Fig. 7) evaluates a model's ability to classify phishing and safe emails, showing strong overall performance. Out of 20 emails, the model correctly identified 11 phishing emails and 6 safe emails. It made only 3 errors – misclassifying 2 safe emails as phishing (false positives) and 1 phishing email as safe (false negative). These results suggest the model is highly accurate, with only minor misclassification issues. The Llama ROC curve (Fig. 8) illustrates the trade-off between the true positive rate and false positive rate for the email classification model. With an AUC (Area Under Curve) of 0.83, the model demonstrates strong discriminative ability – where 1.0 would be perfect

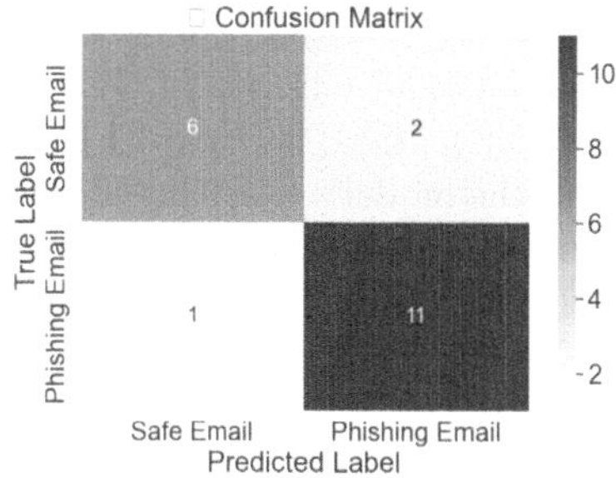

Fig. 7. Confusion matrix for Llama.

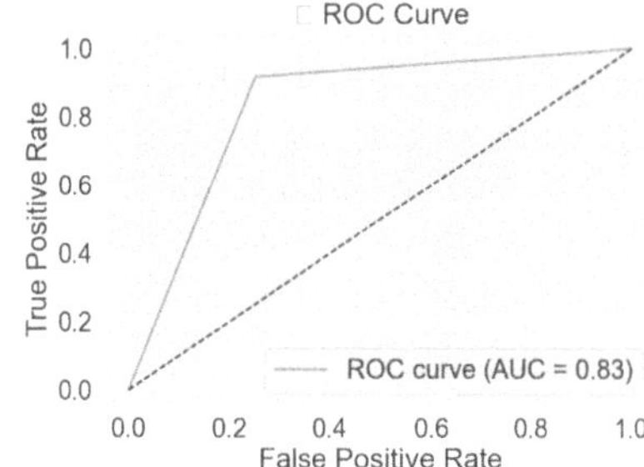

Fig. 8. ROC curve for Llama.

and 0.5 indicates no better than random guessing. The curve's steep rise and high placement suggest the model is effective at distinguishing between phishing and safe emails, although there is still some room for improvement. Finally, the classification report shows that the model achieved an overall accuracy of 85% in detecting phishing and safe emails. For phishing emails, it performed very well with a recall of 0.92 and an F1-score of 0.88, indicating it correctly identified most phishing attempts. For safe emails, it had a slightly lower recall of 0.75, meaning it missed a few, but still maintained a solid precision of 0.86. The macro and weighted averages of the metrics are consistently high (around 0.84–0.85), reinforcing the model's balanced and reliable performance across both classes. One major down side of this solution it is the time elapsed: to evaluate 20 emails, it took 555.06 s.

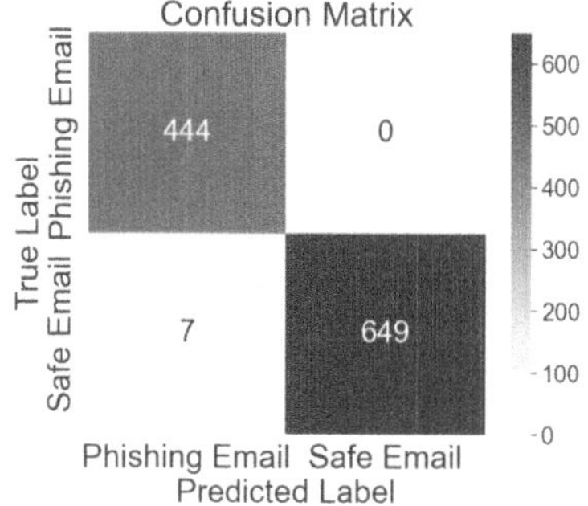

Fig. 9. Confusion matrix for Bayes classifiers.

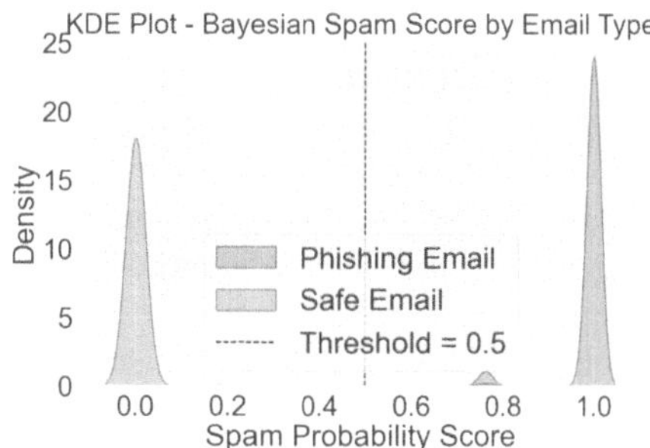

Fig. 10. KDE curve for Bayes classifiers.

The KDE (Kernel Distribution Curve, (Fig. 10)) plot's threshold at 0.5 (dashed line) effectively divides the two distributions, suggesting that the Bayesian model confidently distinguishes phishing from safe emails. The classification report shows that the Bayesian spam filter performed exceptionally well, achieving an overall accuracy of 99% on a test set of 1,100 emails. It correctly identified all 444 phishing emails (100% recall) and 649 out of 656 safe emails, with a perfect precision for safe emails and only 7 false positives (Fig. 9). The high F1-scores of 0.99 for both classes confirm the model's excellent balance between precision and recall. These results indicate that the Bayesian scoring method is highly effective for spam and phishing email detection with minimal error.

Table 1. Key features of the solutions

Property	Hybrid solution	Llama solution	Bayes classifiers solution
Accuracy	0.97	0.85	0.99
ROC curve	0.98	0.83	no
Test emails	1100	20	1100
Elapsed time	33.29	555.06	1.038
Category identifier	yes	no	no

Table 1 summarizes the strengths and limitations of all the evaluated solutions. Based on the presented data, the Bayes classifier emerges as the fastest and most reliable approach. Our hybrid solution ranks second, offering the added advantage of categorizing phishing emails, which provides deeper insight into the nature of each attack. The Llama-based solution ranks last, primarily due to its significant processing time, which limits its practicality for timely email classification.

5 Conclusions

In this study, we proposed a hybrid methodology for phishing email detection that integrates large language models (LLMs) with Bayes classifiers to improve overall detection accuracy and assess the practical applicability of this approach in real-world scenarios. While both techniques were evaluated independently, standalone LLMs exhibited substantial drawbacks, including high computational requirements, lengthy processing times, and comparatively lower accuracy. Conversely, traditional Bayes classifiers, though efficient and previously successful in phishing detection, lack the semantic depth needed to handle more sophisticated attacks. Our hybrid framework effectively combines the contextual understanding of LLMs–despite the overhead of preprocessing and training–with the probabilistic precision of Bayes classifiers, resulting in improved performance across key metrics. Most notably, beyond merely detecting phishing attempts, our approach enables the accurate categorization of phishing email types post-detection, offering a distinct advantage over classical methods and providing deeper insights into the nature of each threat. As a future work, we will like to test the solution also on other available datasets, like: The Enron, Ling, CEAS,

Nazario, Nigerian & Spam Assassin, available on Kaggle Portal [14], under the name *Phishing Email Dataset*, to evaluate the results and metrics. Based upon the results that we obtain, we want to test our solution on a more powerful hardware, more suitable for LLM's.

Acknowledgments. This work has been partially supported by (1) the project RoN-aQCI, part of EuroQCI, DIGITAL-2021-QCI-01- DEPLOY-NATIONAL, 101091562, (2) the project "Romanian Hub for Artificial Intelligence - HRIA", Smart Growth, Digitization and Financial Instruments Program, 2021–2027, MySMIS no. 351416.

References

1. Joshua, G., David, H., Eric, R.: A Bayesian approach to filtering junk e-mail. J. Proc. AAAI Spring Symp. Intell. Agents (2007)
2. Armin, B., Jan, D.B., Sven, G., et al.: A new approach for detecting phishing emails. J. Eur. Symp. Res. Comput. Secur. (ESORICS), 1–20 (2010)
3. Jacob, D., Chang, M.-W., Lee, K., Kristina, T.: BERT: pre-training of deep bidirectional transformers for language understanding. J. Proc. NAACL-HLT, 1–20 (2019). https://doi.org/10.18653/v1/N19-1423
4. Brown, T.B., Mann, B., Ryder, N., et al.: Language models are few-shot learners. In: Journal Advances in Neural Information Processing Systems (NeurIPS) (2020). https://doi.org/10.48550/arXiv.2005.14165
5. Shreshth, A., Mario, F., Christian, R.: Adversarial training for robust phishing detection using BERT. J. Comput. Secur. - ESORICS 2020, 1–20 (2020)
6. Qi, Z., Tao, L., Xiao, W., Jiang, L.: EmailBERT: a pre-trained model for email phishing detection. J. arXiv preprint (2021)
7. Li, H., Singh, P., Patel, K.: Hybrid LLM and bayes classifier for robust phishing detection. In: Journal IEEE Conference on Dependable and Secure Computing (DSC) (2022)
8. Henrik, F., Sam, B., Varun, A., Ben, J., Simon, P.: Devising and detecting phishing emails using large language models. J. IEEE Access (2024). https://doi.org/10.1109/ACCESS.2024.3375882
9. Lee, C.: Enhancing phishing email identification with large language models. J. arXiv preprint (2025)
10. Zihan, J., Wang, P., Lin, J., Tian, D.: Benchmarking and evaluating large language models in phishing detection for small and midsize enterprises: a comprehensive analysis. J. IEEE Access (2025). https://doi.org/10.1109/ACCESS.2025.3540075
11. Hugging Face. https://huggingface.co/. Accessed 16 Apr 2025
12. Zihan, J., Wang, J., Liu, H., Sun, L., et al.: rain small, infer large: memory efficient LoRA training for large language models. In: Journal International Conference on Learning Representations (ICLR (2025)
13. GitHub. https://github.com/. Accessed 25 Apr 2025
14. Kaggle. https://kaggle.com/. Accessed 27 Apr 2025

Special Session: Artificial Intelligence for Protecting the Internet of Things

Influence of Noise on the Stability of a Stochastic SIR Model with Demography

Rafael Rodríguez-García[1], Marta-María Álvarez-Crespo[3], Antonio Díaz-Longueira[3], Carlos Cambra[1], and Roberto Casado-Vara[2(✉)]

[1] Grupo de Inteligencia Computacional Aplicada (GICAP), Departamento de Digitalización, Escuela Politécnica Superior, Av. Cantabria s/n, 09006 Burgos, Spain
{rafael.rodriguez,ccbaseca}@ubu.es

[2] Grupo de Inteligencia Computacional Aplicada (GICAP), Departamento de Matemáticas y Computación, Escuela Politécnica Superior, Av. Cantabria s/n, 09006 Burgos, Spain
rccasado@ubu.es

[3] Universidade de A Coruña, CTC, CITIC, Departamento de Ingeniería Industrial, Rúa Mendizábal s/n, Campus de Esteiro, 15403 Ferrol, A Coruña, Spain
{marta.maria.alvarez.crespo,a.diazl}@udc.es

Abstract. This work analyzes the stability of a stochastic SIR model with demography, where randomness is introduced by adding white noise to the transmission rate. Using a Mean-Square stability framework, we derive a sufficient condition for the stochastic stability of the disease-free equilibrium point and introduce a stochastic basic reproduction number. Numerical simulations confirm that stochasticity alters the epidemic threshold and reduces the number of infections as noise increases, leading to disease extinction even when the deterministic outcome predicts an endemic equilibrium.

Keywords: Stochastic differential equations · Mean-Square stability · SIR model · Noise effects

1 Introduction

Mathematical models are an essential tool for understanding and predicting the spread of infectious diseases. One of the most widely used approaches is deterministic models, which usually divide the population into different compartments and describes the flow between them, being Kermack and McKendrick work [6] pioneering in this field. These global models assume that disease transmission follows smooth, predictable dynamics governed by ordinary differential equations. However, real-world epidemics rarely follow purely deterministic patterns, and

E. Corchado et al. (Eds.): CISIS 2025, CCIS 2807, pp. 137–146, 2026.
https://doi.org/10.1007/978-3-032-19770-2_13

stochastic fluctuations arise because of factors such as demographic noise, environmental variability, and heterogeneity in disease transmission. These sources of uncertainty can significantly alter epidemic outcomes, making it necessary to incorporate stochasticity into mathematical models to capture the inherent randomness of disease spread fully.

One approach, which is used in this study, involves introducing noise into the transmission rate β, leading to a stochastic differential equation (SDE) formulation. Such a model captures variations due to unpredictable external influences and allows us to create a more realistic representation of disease dynamics, particularly in small populations where random effects are significant. Epidemiological models have also been applied in cybersecurity to study the propagation of malware and mitigation strategies, see [4,11]. In particular, stochastic techniques have been incorporated to account for uncertainties in network behavior, [8].

In deterministic models, the concept of stability is analyzed in terms of fixed or equilibrium points, with a basic reproduction number R_0 that determines whether an epidemic will persist or die. However, random fluctuations can destabilize the system, and classical deterministic stability criteria do not fully capture epidemic dynamics. Mean-square stability can be adopted as a rigorous mathematical framework for analyzing stability in stochastic epidemic models leading to the derivation of a stochastic basic reproduction number R_0^S, which serves as a lower bound for the stability of the disease-free equilibrium. Thus, while condition $R_0^S < 1$ provides a sufficient condition for the MS- stability of the disease free equilibrium, it does not necessarily define the true epidemic threshold, as extinction can be reached even when $R_0^S > 1$. Numerical results by Tornatore et al. [10] suggest the existence of an upper bound where system becomes unstable and oscillates around the endemic equilibrium, highlighting that considering noise alters the threshold for an epidemic to occur.

In summary, while previous studies [9] have derived sufficient conditions for Mean-Square stability in stochastic epidemiological models, they have not fully explored how noise influences extinction probability. This study addresses this gap by demonstrating numerically that increasing noise intensity can suppress outbreaks that would otherwise persist in a deterministic formulation.

This paper is organized as follows: Sect. 2 introduces the mathematical background and stability criteria. In Sect. 3, we derive the stochastic reproduction number and analyze its dependence on noise. Section 4 presents the numerical simulations that illustrate our theoretical findings. Finally, Sect. 5 concludes the study and discusses future research directions.

2 Preliminaries

A stochastic differential equation (SDE) is an extension of an ordinary differential equation (ODE) with a random perturbation term. A general SDE is given by:

$$dX_t = f(X_t, t)dt + g(X_t, t)dW_t \tag{1}$$

Here, W_t is a Wiener process (or Brownian motion), which models continuous random fluctuations with independent normally distributed increments [2]. The integration is understood in the Itô sense [12], which is standard for most stochastic modelling applications.

The stochastic stability behavior of a stochastic differential equation can be derived from the study of its linearized equation. See [7,9] for further details. In the case of this study, where there is only one source of noise (scalar noise):

$$dX_t = AX_t dt + BX_t dW_t \tag{2}$$

where A and B are constant matrices evaluated in the fixed point.

$$A = \frac{\partial f}{\partial x}(0) \quad B - \frac{\partial g}{\partial x}(0) \tag{3}$$

Second moment $P(t) = \mathbb{E}[X_t X_t'] = (p_{ij}(t))$ satisfies the following equation:

$$\frac{dP(t)}{dt} = AP(t) + P(t)A' + BP(t)B', \tag{4}$$

Following Arnold [2], Mean-Square (MS) stability at the equilibrium point is equivalent to the trivial solution of system 4. Reordering terms in (4) and noting that $P(t)$ is symmetric, we can rewrite the system as $d(d+1)/2$ differential equations:

$$\frac{dY}{dt} = \mathcal{M}Y. \tag{5}$$

where the elements of Y correspond to the distinct entries of $P(t)$, taking into account its symmetry. In other words, Y contains only the unique elements of $P(t)$, since $p_{ij} = p_{ji}$ by definition. $\mathcal{M}$ is the linear operator governing the dynamics of the second moment vector. For example, in the three-dimensional system considered in this study:

$$Y = (p_{11}(t), p_{22}(t), p_{33}(t), p_{12}(t), p_{13}(t), p_{23}(t))^T \tag{6}$$

Therefore, MS-stability is equivalent to ordinary stability at the equilibrium point $Y = 0$ of 5. Denoting by $\sigma(\mathcal{M})$ the spectrum of the matrix $\mathcal{M}$ and:

$$\nu(\mathcal{M}) := \max\{\Re(\lambda) : \lambda \in \sigma(\mathcal{M})\} \tag{7}$$

Its spectral abscissa,i.e. the largest real part among all eigenvalues of $\mathcal{M}$, determining the exponential growth or decay of the second moment system. We arrive at the following criterion:

Proposition: The linear system 2 is asymptotically mean-square stable if and only if $\nu(\mathcal{M}) < 0$.

3 Epidemiological Models

The classical SIR model describes the spread of infectious diseases by dividing the population into three compartments: susceptible (S), infected(I), and recovered (R). In its standard form, it assumes a closed population with no births or deaths, which is called an epidemic model without explicit demography. However, turnover within the population due to the addition of newborns and the removal of individuals through death can create new scenarios, such as the persistence of the infection. To account for this, we introduce a constant birth and death rate μ to formulate the SIR model with demography

$$\begin{cases} S'(t) = (-\beta SI - \mu S + \mu) \\ I'(t) = (\beta SI - \gamma I) \\ R'(t) = (\gamma I - \mu R) \end{cases} \tag{8}$$

where:

- μ is the birth and death rate (all newborns become susceptible)
- β is the transmission rate
- γ is the recovery rate

Note that $S'(t) + I'(t) + R'(t) = 0$, therefore the system assumes a constant and normalized population, i.e. $S + I + R = 1$. This extension leads us to the basic reproduction number:

$$R_0 = \frac{\beta}{\gamma + \mu} \tag{9}$$

The long-term behaviour depends on R_0:

- If $R_0 < 1$ the infection dies out ($\lim_{t\to\infty} I(t) = 0$), leading to the disease free equilibrium $(1, 0, 0)$, which is stable: any initial infection will eventually disappear.
- If $R_0 > 1$, the disease free equilibrium is unstable, any $I(0) > 0$ will drive the system toward the endemic equilibrium.

$$S^* = \frac{\gamma + \mu}{\beta} \quad I^* = \frac{\mu(\beta - \gamma - \mu)}{\beta(\mu + \gamma)} \quad R^* = \frac{\gamma(\beta - \gamma - \mu)}{\beta(\mu + \gamma)} \tag{10}$$

Unlike the classical SIR model, incorporating demography allows for both possible outcomes: disease extinction or persistence, depending on the parameter values.

3.1 Stochastic Model

The deterministic approach using the compartmental SIR model is the classical framework. However, due to the inherent randomness of epidemiological processes, it is often more appropriate to resort to stochastic techniques. In particular, the infection rate β is highly variable. Therefore, it becomes interesting to propose compartmental models governed by *stochastic differential equations*, where the parameter β is affected by the addition of *white noise* ξ_t, with intensity σ, resulting in $\beta + \sigma\xi_t$.

In this way, the deterministic SIR model transforms into a stochastic model in the sense of Itô. Using the normalized system in 8 as a reference, we analyze how the associated parameters are modified. Specifically, the basic reproduction number, which is traditionally defined in terms of the infection rate, must be adapted accordingly. This leads to the following modification:

$$\begin{cases} dS = (-\beta SI - \mu S + \mu)dt - \sigma SIdW \\ dI = (\beta SI - \gamma I)dt + \sigma SIdW \\ dR = (\gamma I - \mu R)dt \end{cases} \tag{11}$$

where W_t is a Wiener process and σ is the intensity of the stochastic perturbation (noise strength). The linearized matrix evaluated in the disease free fixed point for this system:

$$A = \begin{pmatrix} -\mu & -\beta & 0 \\ 0 & \beta-\gamma-\mu & 0 \\ 0 & \gamma & -\mu \end{pmatrix} \qquad B = \begin{pmatrix} 0 & -\sigma & 0 \\ 0 & \sigma & 0 \\ 0 & 0 & 0 \end{pmatrix} \tag{12}$$

From 4 can be obtained a 6×6 matrix:

$$\mathcal{M} = \begin{pmatrix} -2\mu & \sigma^2 & 0 & -2\beta & 0 & 0 \\ 0 & 2(\beta-\gamma-\mu)+\sigma^2 & 0 & 0 & 0 & 0 \\ 0 & 0 & -2\mu & 0 & 0 & 2\gamma \\ 0 & -\beta-\sigma^2 & 0 & \beta-\gamma-2\mu & 0 & 0 \\ 0 & 0 & 0 & \gamma & -2\mu & -\beta \\ 0 & \gamma & 0 & 0 & 0 & \beta-\gamma-2\mu \end{pmatrix} \tag{13}$$

Due to the sparsity and structure of M, the eigenvalue calculation yields the diagonal elements.

$$\begin{aligned} &\lambda_1 = -2\mu, \quad \lambda_2 = 2(\beta-\gamma-\mu)+\sigma^2, \quad \lambda_3 = -2\mu, \\ &\lambda_4 = \beta-\gamma-2\mu, \quad \lambda_5 = -2\mu, \quad \lambda_6 = \beta-\gamma-2\mu \end{aligned} \tag{14}$$

Since all parameters in the simulation are positive, we have $\lambda_1, \lambda_3, \lambda_5 < 0$. Now, if $\lambda_2 < 0$, we obtain the following condition:

$$\beta - \gamma - \mu < 0$$

Thus, $\lambda_4, \lambda_6 < 0$. In other words, to achieve MS-stability, we must impose 7:

$$\lambda_2 < 0 \to 2(\beta-\gamma-\mu)+\sigma^2 < 0 \to \beta < \gamma + \mu - \frac{\sigma^2}{2} \tag{15}$$

This condition coincides with the sufficient stability condition found by Tornatore et al. An alternative expression can be obtained using deterministic R_0:

$$R_0 + \frac{\sigma^2}{2(\gamma+\mu)} < 1 \tag{16}$$

An stochastic basic reproductive number can be defined now:

$$R_0^s = R_0 + \frac{\sigma^2}{2(\gamma + \mu)} \tag{17}$$

This expression indicates that the stochastic reproductive number increases with noise intensity compared to the deterministic case. Therefore, the presence of stochastic fluctuations makes the threshold for disease persistence higher, increasing the probability of disease extinction and reducing the mean number of infections. This phenomenon is known as noise-induced extinction or stochastic stabilisation [1]

Disease free equilibrium remains MS-stable under the conditions previously established. Aditionally, Tornatore et al. [10] found numerically that these are sufficient conditions for stability, as the system remains asymptotically stable when β satisfies the following constraint:

$$\min\{\gamma + \mu - \frac{\sigma^2}{2}, 2\mu\} < \beta < \gamma + \mu + \frac{\sigma^2}{2} \tag{18}$$

Conversely, when β exceeds the upper bound, system becomes unstable, leading solution to oscillate around the endemic equilibrium.

4 Simulations and Numerical Results

In this section, we present the results obtained from the simulations of the SIR model with demography. Stochastic and deterministic trajectories are compared for different values of the noise intensity σ, stability conditions are also analyzed for the disesase free equilibrium point.

Stochastic system (Eq. 11) is solved numerically using Euler- Maruyama method, while deterministic trajectories from Eq. 8 are computed using the Runge-Kutta method of order 4. Both methods are standard numerical approaches for solving differential equations, and their stability properties are well studied in the literature [3,5]. All data presented in this section are synthetic, generated by numerical integration, with each trajectory consisting of 501 time steps ($t \in [0, 500]$ with step size $h = 1/10$).

We performed two groups of experiments and the following parameters are fixed for all simulations:

$$\gamma = 0.1 \quad \mu = 0.2 \quad S_0 = 0.975 \quad I_0 = 0.025 \quad R_0 = 0 \tag{19}$$

where (S_0, I_0, R_0) are normalized initial conditions. These parameter values are typical for epidemiological simulations, with a small initial fraction of infected individuals to test disease spread. They were also chosen strategically to cross the stability threshold as σ increases, as will be shown below.

For the first experiment: $\beta = 0.37$, satisfies condition $\beta > \gamma + \mu$, i.e. $R_0 > 1$, which implies that deterministic trajectory will converge to the endemic equilibrium point.However, in the stochastic model, the occurrence of an epidemic depends on a new threshold influenced by σ. From 18,we redefine the lower bound as $C_1 = \gamma + \mu - \frac{\sigma^2}{2}$ and upper bound as $C_2 = \gamma + \mu + \frac{\sigma^2}{2}$. In figures (a) and (b) from 1, we observe that $\beta > C_2$, which implies an unstable system oscillating around endemic equilibrium point. Aditionally, in figure (b) $\sigma = 0.2$, oscillations become stronger compared to case (a), highlighting the increasing effect of noise in the dynamics. On the other hand, Figures (c) and (d) from 1, where $\beta < C_2$, shows that infection can die out even when deterministic outcome is an endemic equilibrium.

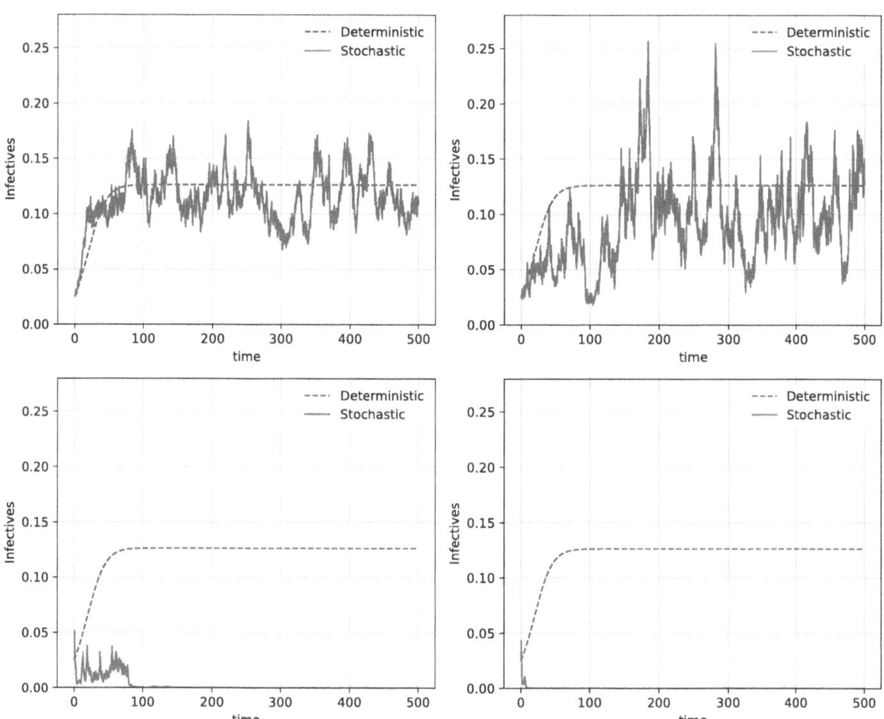

Fig. 1. Time evolution of infectives for different values of sigma; (a) $\sigma = 0.1$ (top left), (b) $\sigma = 0.2$ (top right), (c) $\sigma = 0.4$ (bottom left), (d) $\sigma = 0.6$ (bottom right).

The results from the stochastic simulations indicate that the presence of noise in the system tends to drive the infection toward extinction. This effect becomes evident in the Fig. 2, where we average over 10000 stochastic trajectories from experiment 1 of the Fig. 1:

$$\mathbb{E}[I] = \frac{1}{10000} \sum_{j=1}^{10000} I_j \tag{20}$$

For second experiment we set $\beta = 0.27$, therefore $\beta < C_2 < \gamma + \mu$ and infection is expected to die in every situation as the sufficient condition for MS-stability is supplied. We can observe that in Fig. 3, where trajectories are averaged over 10000 runs.

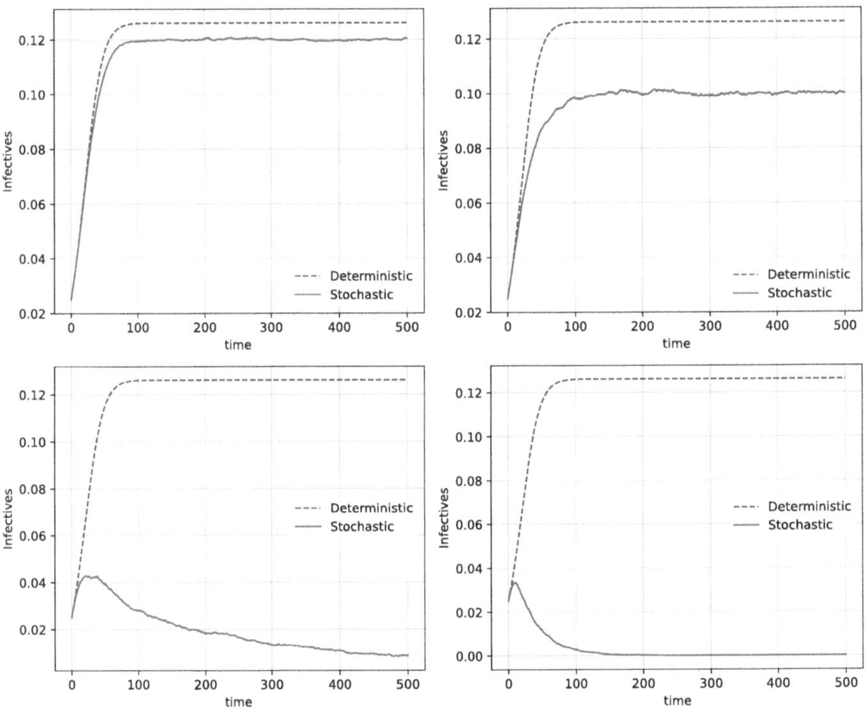

Fig. 2. Time evolution of the mean infected population($\beta = 0.37$), averaged over 10000 trajectories for different values of σ (a) $\sigma = 0.1$ (top left), (b) $\sigma = 0.2$ (top right), (c) $\sigma = 0.4$ (bottom left) , (d) $\sigma = 0.6$ (bottom right).

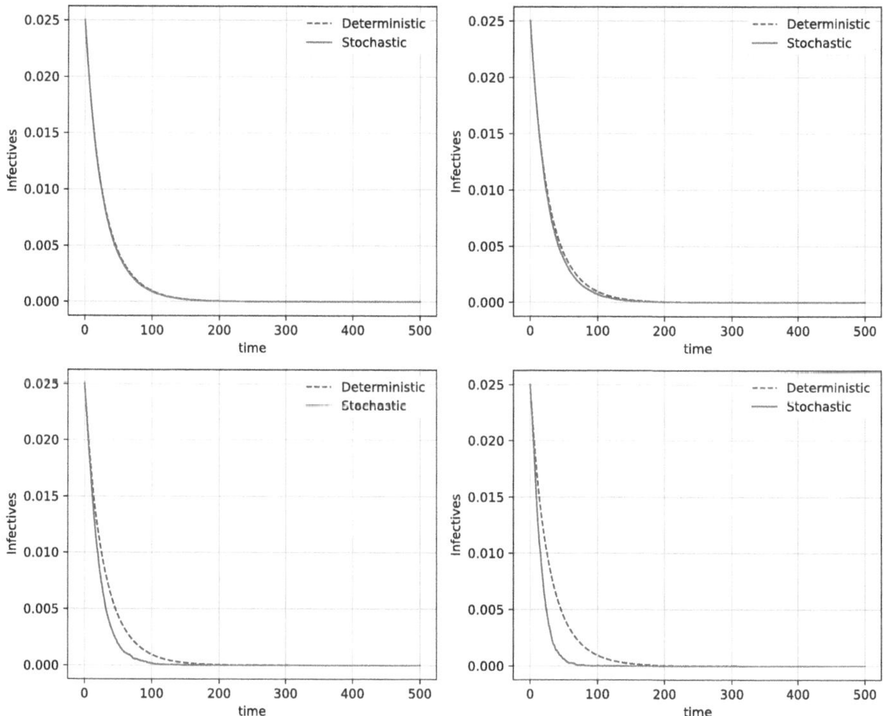

Fig. 3. Time evolution of the mean infected population($\beta = 0.27$), averaged over 10000 trajectories for different values of σ (a) $\sigma = 0.1$ (top left), (b) $\sigma = 0.2$ (top right), (c) $\sigma = 0.4$ (bottom left) , (d) $\sigma = 0.6$ (bottom right).

5 Conclusions and Future Work

This study analyzes the impact of stochastic fluctuations on the stability of the SIR model with demography. Using a Mean-Square stability framework, we established a sufficient condition ($R_0^s < 1$) for stochastic stability. Stochastic perturbations in the transmission rate modify the model, creating a new threshold that determines whether the disease persists or becomes extinct.

Our numerical simulations, particularly those shown in Fig. 2, confirm these theoretical results. They demonstrate that including noise reduces the mean number of infections and can induce disease extinction even when deterministic models predict persistence. This highlights the importance of considering stochastic effects when assessing epidemic thresholds and planning intervention strategies.

Acknowledgments. This publication is part of the AI4SECIoT project ("Artificial Intelligence for Securing IoT Devices"), funded by the National Cybersecurity Institute (INCIBE), derived from a collaboration agreement signed between the National Institute of Cybersecurity (INCIBE) and the University of Burgos. This initiative is carried out within the framework of the Recovery, Transformation and Resilience Plan funds, financed by the European Union (Next Generation), the project of the Government of Spain that outlines the roadmap for the modernization of the Spanish economy, the recovery of economic growth and job creation, for solid, inclusive and resilient economic reconstruction after the COVID19 crisis, and to respond to the challenges of the next decade.

References

1. Appleby, J.A., Mao, X., Rodkina, A.: Stabilization and destabilization of nonlinear differential equations by noise. IEEE Trans. Autom. Control **53**(3), 683–691 (2008)
2. Arnold, L.: Stochastic Differential Equations: Theory and Applications. John Wiley & Sons, New York (1974)
3. Butcher, J.C.: Numerical Methods for Ordinary Differential Equations. John Wiley & Sons (2016)
4. Feng, L., Liao, X., Han, Q., Li, H.: Dynamical analysis and control strategies on malware propagation model. Appl. Math. Model. **37**(16–17), 8225–8236 (2013)
5. Higham, D.J.: An algorithmic introduction to numerical simulation of stochastic differential equations. SIAM Rev. **43**(3), 525–546 (2001)
6. Kermack, W., McKendrick, A.: Contributions to the mathematical theory of epidemics, part i. Proc. Roy. Soc. Lond. Ser. A **115**, 700–721 (1927)
7. Khasminskii, R.: Stochastic Stability of Differential Equations, 2nd edn. Springer, Berlin (2012)
8. Mahboubi, A., Camtepe, S., Ansari, K.: Stochastic modeling of IoT botnet spread: a short survey on mobile malware spread modeling. IEEE Access **8**, 228818–228830 (2020)
9. Tocino, A., Senosiain, M.J.: Mean-square stability analysis of numerical schemes for stochastic differential systems. J. Comput. Appl. Math. **236**(10), 2660–2672 (2012)
10. Tornatore, E., Buccellato, S.M., Vetro, P.: Stability of a stochastic SIR system. XXPhys. A **354**, 111–126 (2005)
11. Yu, S., Gu, G., Barnawi, A., Guo, S., Stojmenovic, I.: Malware propagation in large-scale networks. IEEE Trans. Knowl. Data Eng. **27**(1), 170–179 (2014)
12. Øksendal, B.: Stochastic Differential Equations: An Introduction with Applications. Springer Science & Business Media, Cham (2010)

An Optimal Control Problem for a SIR Model with Two Mitigation Strategies for Malware Spread

Roberto Casado-Vara[1(✉)], Rafael Rodríguez García[2], Carlos Cambra[2], Agustín García-Fischer[3], Esteban Jove[3], and Alvaro Herrero[2]

[1] Grupo de Inteligencia Computacional Aplicada (GICAP), Departamento de Matemáticas y Computación, Escuela Politécnica Superior, Universidad de Burgos, Av. Cantabria s/n, 09006 Burgos, Spain
rccasado@ubu.es
[2] Grupo de Inteligencia Computacional Aplicada (GICAP), Departamento de Digitalización, Escuela Politécnica Superior, Universidad de Burgos, Av. Cantabria s/n, 09006 Burgos, Spain
{rafael.rodriguez,ccbaseca,ahcosio}@ubu.es
[3] Departamento de Ingeniería Industrial, Universidade de A Coruña, CTC, CITIC, Rúa Mendizábal s/n, Campus de Esteiro, 15403 Ferrol, A Coruña, Spain
{agustin.garciaf,esteban.jove}@udc.es

Abstract. The increasing complexity of IoT networks heightens the risk of malware propagation, requiring efficient mitigation strategies. Traditional cybersecurity solutions are often reactive and do not optimize resource allocation for malware containment. This study formulates an optimal control problem for malware spread using a Susceptible-Infected-Recovered (SIR) model, combining two strategies: reducing transmission and accelerating removal. The control problem is modeled with the Hamilton-Jacobi-Bellman (HJB) equation and solved numerically. The results show that optimal preventive and remedial actions vary with network conditions: high infection levels call for stronger prevention, while widespread infections require intensive remediation. Compared to conventional approaches, this method improves cost efficiency and resource allocation, demonstrating the scalability and effectiveness of HJB-based strategies for IoT cybersecurity.

Keywords: IoT security · malware propagation · optimal control · HJB equation · SIR model

1 Instroduction

The spread of malware in IoT networks is a significant cybersecurity issue, especially given the exponential growth of connected devices [1]. Epidemiological models such as SIR have been used to describe infection dynamics in vulnerable devices [2]. However, current approaches are often reactive and do not optimise the allocation of resources to efficiently prevent and remove malware. One

E. Corchado et al. (Eds.): CISIS 2025, CCIS 2807, pp. 147–156, 2026.
https://doi.org/10.1007/978-3-032-19770-2_14

of the key challenges is to design an optimal and dynamic strategy that minimizes both the number of infected devices and the cost of intervention. Existing solutions, such as firewalls and antivirus, are limited because they do not incorporate a mathematical approach to determine when and where to apply control measures [3,4]. In addition, traditional malware detection and removal methods usually require manual monitoring and do not provide real-time response [5]. This study approaches the problem from an optimal control approach based on the HJB equation [6]. A strategy is proposed that combines two mitigation measures: reducing the spread of malware by restricting device communication and accelerating malware removal through security updates and automated tools. Using a SIR model adapted to IoT networks, controls are defined that regulate the transmission rate and recovery rate of malware. The aim is to develop a mathematical approach for optimizing the response to malware attacks in IoT networks. We hypothesize that an optimal control strategy based on the HJB equation can significantly improve the efficiency of malware mitigation compared to conventional methods. To solve this problem, we model the spread of malware using a system of differential equations and formulate an optimal control problem with two control variables: one representing the effort to reduce the spread of malware and the other representing the intensity of removal actions on infected devices. The resulting HJB equation is solved numerically by finite difference, discretising the state space into a network that allows us to obtain optimal policies as a function of the number of susceptible and infected devices.

The results show that the optimal mitigation strategy varies according to the state of the IoT network [5]. The value function obtained indicates that high infection scenarios require more costly interventions, while in low infection states the cost is lower. The optimal policy to reduce the spread of malware is triggered when there is a high number of susceptible and infected nodes, suggesting that aggressive prevention is the best strategy. In contrast, when infection is already widespread, the priority shifts to removing malware through security scans and updates. These findings confirm that a strategy based on optimal control allows an efficient allocation of resources, reducing both the number of infected devices and the cost of intervention. Compared to traditional approaches, this method improves the response to attacks, optimizes the allocation of security measures and provides a solution that is adaptable to different cyber security environments [5,7]. While the SIR model is useful for providing insights into malware mitigation, we acknowledge its limitations, such as oversimplifying network structure and assuming permanent immunity. Nevertheless, it offers a practical starting point for developing and analysing optimal control strategies. Unlike traditional approaches that rely on fixed or reactive mitigation, our method formulates malware containment as a dynamic optimal control problem using the HJB equation. This enables preventive and remedial actions to be optimised jointly and adaptively according to the network's current state, ensuring efficient resource allocation and minimising intervention costs. This constitutes a significant improvement over existing methods. This paper is organized as follows: we present the mathematical background and the proposed optimal control problem in Sect. 2. Section 3 presents the setup of our simulations and the results which validate the performance of the proposal. Finally, Sect. 4 concludes the conducted research and proposes future lines of work.

2 Mathematical Model and Optimal Control Problem

We explore the SIR epidemic model under the assumption of a constant total amount of IoT devices (i.e., population size). The IoT devices are categorised into three distinct compartments based on disease status: susceptible devices (S), which are at risk of contracting the infection; infectious devices (I), which are actively infected and capable of spreading the disease; and recovered devices (R), which have overcome the infection and acquired lifelong immunity. We assume that transmission occurs exclusively through direct contact with infectious devices and that once recovered, devices remain permanently immune since we assume that security patches are truly effective against malware. The following parameters govern transitions between these compartments: $S(t)$ represents the susceptible IoT devices, $I(t)$ represents the infectious IoT devices, $R(t)$ represents the recovered IoT devices, β is the effective contact rate and γ is the recovery rate. The control strategy we adopt in this work consists of the simultaneous implementation of two malware mitigation measures in a IoT network:

- Reduction of malware propagation by restricting contact between susceptible and infected nodes.
- Acceleration of malware removal in infected nodes through the application of security tools.

The aim is to minimize the number of infected IoT devices and reduce the cost associated with interventions. To achieve this, we introduce two control functions, $u_1(t)$ and $u_2(t)$ into the model:

- $u_1(t)$ represents the fraction of susceptible devices that apply protective measures, such as firewalls, network segmentation, or advanced security protocols, to reduce malware transmission.
- $u_2(t)$ represents the fraction of infected devices receiving active interventions, such as antivirus deployment, security updates, or system restorations, with the goal of eliminating malware and restoring the device's functionality.

To model this control strategy, we modify the differential equations of the classical SIR model for malware propagation in networks:

$$\begin{cases} \dfrac{dS}{dt} = -\beta(1-u_1)SI, \\ \dfrac{dI}{dt} = \beta(1-u_1)SI - (\gamma + u_2)I, \\ \dfrac{dR}{dt} = (\gamma + u_2)I. \end{cases} \tag{1}$$

2.1 The Optimal Control Problem

Our goal is to reduce the number of susceptible and infected IoT devices, while maximizing the number of recovered devices over the course of the uncontrolled malware propagation (i.e., epidemic). Mathematically, for a given terminal time T, the problem is to minimize the following objective function:

$$J(u_1, u_2) = \int_0^T \left[AI + Bu_1^2 + Cu_2^2\right] dt, \tag{2}$$

where:

- AI penalizes the number of infected devices in the network.
- Bu_1^2 represents the cost of implementing measures to reduce malware spread, which may include network connectivity restrictions, segmentation, or firewalls.
- Cu_2^2 accounts for the cost of applying malware removal tools, such as security updates and antivirus scans.

To ensure that the control actions are well-defined and feasible within a real-world cybersecurity framework, we impose constraints on the set of admissible control functions. Specifically, the applied mitigation and recovery strategies should remain bounded and measurable over the entire time horizon. Thus, we define the set of admissible controls defined by

$$U = \left\{(u_1, u_2) \in L^\infty([0,T]) \times L^\infty([0,T]) \;\middle|\; 0 \le u_1(t) \le u_{1,\max},\ 0 \le u_2(t) \le u_{2,\max},\ \forall t \in [0,T]\right\}. \tag{3}$$

2.2 Existence of an Optimal Control

In this section, we establish the existence of an optimal control for the malware propagation problem using standard results from optimal control theory by applying the procedure and several results from Fleming and Rishel in [8]. For that purpose we will first define when a solution of the HJB equation is a viscosity solution of the HJB equation using the definition of Fabbri adapted to our optimal control problem [16].

Theorem 1. *Consider the optimal control problem defined by the system dynamics shown in equation (1) and the objective function shown in Eq. (2). If the state domain (S, I) is compact, the controls u_1, u_2 are restricted in $[0, u_{\max}]$, and the value function $V(S, I, t)$ is Lipschitz continuous, then:*

1. *There exists a solution $V(S, I, t)$ to the Hamilton–Jacobi–Bellman (HJB) equation in the viscosity sense.*
2. *There exists at least one pair of optimal controls (u_1^*, u_2^*) that minimize $J(u_1, u_2)$*

Proof. Three steps lead to the existence of the optimal solution:

Step 1: Existence of the Value Function V(S, I, t)

We define the value function as:

$$V(S, I, t) = \min_{u_1, u_2} J(u_1, u_2). \tag{4}$$

To ensure the existence of V(S, I, t) , we verify that:

- $J(u_1, u_2)$ is lower bounded since the terms AI, Bu_1^2, Cu_2^2 are positive.
- $J(u_1, u_2)$ is convex in the controls u_1, u_2 due to the quadratic nature of $Bu_1^2 + Cu_2^2$.
- The domain $\Omega = \{(S, I) \in [0, 1] \times [0, 1] \mid S + I \leq 1\}$ is compact, which allows us to apply Weierstrass' Theorem to guarantee the existence of a continuous value function.

Thus, $V(S, I, t)$ exists as a continuous function that satisfies the HJB equation.

Step 2: Existence of Optimal Controls in a Compact Set:

The optimal controls u_1^*, u_2^* must minimize the Hamiltonian:

$$H = AI + Bu_1^2 + Cu_2^2 + V_S f_S + V_I f_I. \tag{5}$$

Since u_1, u_2 belongs to the compact interval $[0, u_{\max}]$, the function H is defined over a compact domain. By Weierstrass' Theorem in optimal control problems [10], every continuous function attains its minimum in a compact set. Since the Hamiltonian is continuous in u_1, u_2, there exists at least one optimal pair (u_1^*, u_2^*) that minimizes the Hamiltonian H.

Step 3: Proving that the value function is a viscosity solution of the HJB equation:

To ensure the existence of a solution to the HJB equation, we consider the definition of viscosity solutions given in [16]. Suppose that $\phi(S, I, t)$ is a differentiable test function and that $V(S, I, t) - \phi(S, I, t)$ attains a local maximum at (S_0, I_0, t_0), that is:

$$V(S_0, I_0, t_0) = \phi(S_0, I_0, t_0). \tag{6}$$

Taking the time derivative:

$$\frac{\partial V}{\partial t}(S_0, I_0, t_0) \leq \frac{\partial \phi}{\partial t}(S_0, I_0, t_0). \tag{7}$$

Replace in the HJB equation:

$$-\frac{\partial \phi}{\partial t} \geq \min_{u_1, u_2} \left[AI + Bu_1^2 + Cu_2^2 + V_S f_S + V_I f_I\right]. \tag{8}$$

Since ϕ is a local approximation of V at (S_0, I_0, t_0), and the Hamiltonian is convex in the controls, it follows that V satisfies the inequality in the viscosity sense. This proves that V is a super-solution of the HJB equation. A similar argument can be used to prove the sub-solution case.

Since we have proven the existence of $V(S, I, t)$ and that the optimal controls exist in a compact set, it follows from [9] that:

$$-\frac{\partial V}{\partial t} = \min_{u_1, u_2} H(S, I, u_1, u_2, V_S, V_I). \tag{9}$$

Thus, there exists a viscosity solution to the HJB equation, proving the existence of the optimal solution. □

2.3 Characterization of the Optimal Control

This section presents the optimal control theorem, which characterizes the optimal mitigation strategy in terms of explicit control policies. We then provide proof using the HJB equations to ensure that the derived controls effectively minimize the cost function while limiting malware propagation.

Theorem 2. *Consider the optimal control problem defined by the system dynamics shown in Eq. (1) and the objective function shown in Eq. (2). Then, the optimal controls that minimize $J(u_1, u_2)$ are given by:*

$$u_1^*(t, S, I) = \min\left\{u_{1,\max}, \max\left\{0, \frac{\beta SI(V_S - V_I)}{2B}\right\}\right\} \tag{10}$$

$$u_2^*(t, S, I) = \min\left\{u_{2,\max}, \max\left\{0, \frac{IV_I}{2C}\right\}\right\}. \tag{11}$$

where $V(S, I, t)$ satisfies the HJB equation:

$$-\frac{\partial V}{\partial t} = \min_{u_1, u_2}\left[AI + Bu_1^2 + Cu_2^2 + V_S f_S + V_I f_I\right], \tag{12}$$

with the final condition $V(T, S, I) = 0$.

Proof. The proof of the theorem is in three parts: The value function $V(S, I, t)$ represents the optimal future cost from the current state (S, I). Applying Bellman's Optimality Principle [11], the HJB equation is:

$$-\frac{\partial V}{\partial t} = \min_{u_1, u_2}\left[AI + Bu_1^2 + Cu_2^2 + V_S f_S + V_I f_I\right], \tag{13}$$

where:

$$f_S = -\beta(1 - u_1)SI, \quad f_I = \beta(1 - u_1)SI - (\gamma + u_2)I. \tag{14}$$

Substituting f_S and f_I into the HJB equation:

$$-\frac{\partial V}{\partial t} = \min_{u_1, u_2}\left[AI + Bu_1^2 + Cu_2^2 + V_S(-\beta(1 - u_1)SI) + V_I(\beta(1 - u_1)SI - (\gamma + u_2)I)\right]. \tag{15}$$

To minimize the equation, we differentiate for u_1 and u_2: For u_1:

$$\frac{\partial}{\partial u_1}\left(Bu_1^2 + V_S(-\beta(1 - u_1)SI) + V_I(\beta(1 - u_1)SI)\right) = 0. \tag{16}$$

Expanding:

$$2Bu_1 + \beta SI(V_S - V_I) = 0. \tag{17}$$

Solving for u_1^*:

$$u_1^* = \frac{\beta SI(V_S - V_I)}{2B}. \tag{18}$$

Since $u_1 \in [0, u_{1,\max}]$, we impose constraints:

$$u_1^*(t, S, I) = \min\left\{u_{1,\max}, \max\left\{0, \frac{\beta SI(V_S - V_I)}{2B}\right\}\right\}. \tag{19}$$

The same method is used to calculate u_2^*, taking into account its influence on H and solving the corresponding equation. Since u_1^* and u_2^* were determined by minimizing the equation, they provide the optimal solution to the control problem. □

3 Results

3.1 Simulation Setup

The numerical simulations are based on the discretized version of the HJB equation, which is solved using discrete differences in the domain $\Omega = \{(S, I) \in [0, 1] \times [0, 1] \mid S + I \leq 1\}$. The computational domain is defined over the state space (S, I), representing the fraction of susceptible and infected nodes in the IoT network. The model parameters used in the simulations are inspired by the parameters in [14,15] with slight modifications to fit the specific objectives of this study. The chosen parameters are $\beta = 0.5$, $\gamma = 0.1$, $A = 1.0$, $B = 0.01$, and $C = 0.01$ and $u_1 \in [0, 1]$, $u_2 \in [0, 1]$. The computational domain is discretized using $N_s = 50$ and $N_i = 50$ grid points for the susceptible and infected state variables. The time horizon is set to $T = 1.0$ with $N_t = 50$ time steps. The backward Euler scheme ensures stability in the numerical integration.

3.2 Simulation Results

Figure 1 shows the value function $V(S, I)$, which represents the optimal accumulated cost starting from each initial state (S, I). The domain is restricted to the physically meaningful region $S + I \leq 1$, corresponding to feasible states of the SIR model where $R = 1 - S - I \geq 0$. Higher values of V occur in states with a larger proportion of infected nodes, indicating that large-scale infections require more costly interventions. In contrast, states with fewer infected nodes exhibit lower values of V, suggesting that less intervention is needed and the overall cost remains low.

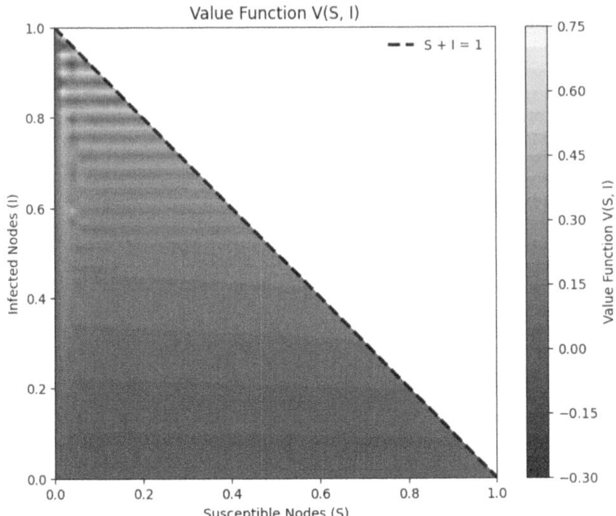

Fig. 1. Value function $V(S, I)$. The function represents the optimal accumulated cost from any initial state. Higher values indicate states with high infection levels requiring costly interventions.

Figure 2 (left) shows the optimal preventive control $u_1^*(S, I)$, which reduces malware transmission among susceptible nodes. The control intensity is highest near the upper region of the feasible domain, where the proportion of infected nodes is large but a significant number of susceptible nodes remain available for infection. In these conditions, strong preventive actions are optimal to slow down the spread. As the number of susceptible nodes decreases (i.e., near the diagonal boundary $S + I = 1$), u_1^* drops sharply, reflecting that prevention becomes less relevant once few nodes remain at risk. Figure 2 (right) depicts the optimal removal control $u_2^*(S, I)$, associated with the recovery or disinfection of infected nodes. The control reaches its highest values along the lower and right boundaries of the feasible domain, corresponding to states where the infection level is moderate but many susceptible nodes remain available for reinfection. In these regions, a strong remediation effort is optimal to prevent further spread by rapidly eliminating existing infections. In contrast, the upper-left part of the triangle shows alternating patches of high and low control intensity, reflecting a balance between containment and diminishing returns once the infection saturates the network.

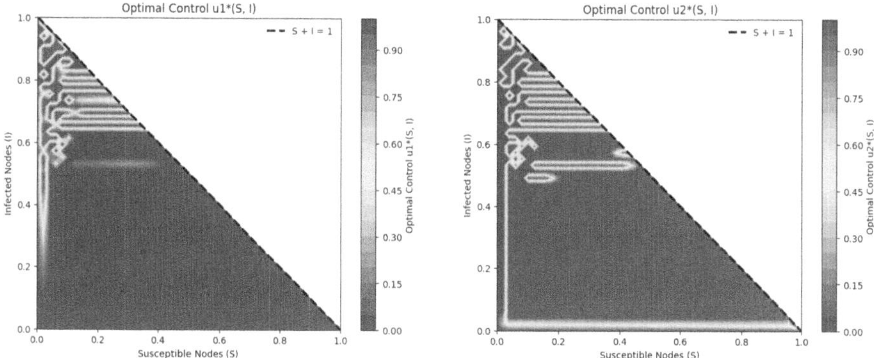

Fig. 2. Optimal control policies for malware propagation mitigation in the domain Ω. Panel (left) shows the optimal control $u_1^*(S, I)$, which represents the intensity of preventive measures aimed at reducing malware transmission. Panel (right) displays the optimal control $u_2^*(S, I)$, associated with the removal or recovery of infected nodes.

4 Conclusion

This study presents an optimal control framework for mitigating malware propagation in IoT networks using an HJB-based approach. By integrating two control strategies - reducing malware transmission and accelerating its removal - the model dynamically adapts interventions based on network conditions. The results indicate that high infection scenarios require aggressive prevention measures, while widespread infections require targeted remediation efforts. Compared to conventional cybersecurity strategies, this approach optimises resource allocation and reduces intervention costs. The proposed methodology increases the efficiency of malware containment, providing a scalable and adaptable solution for IoT security. Future work can extend this model to incorporate stochastic dynamics and network topology variations to improve real-world applicability. This work is a proof of concept based on simulations of a classical SIR model; it does not use real-world IoT data. Future work will involve validating the proposed approach using real datasets and larger-scale simulations, in order to assess its practical applicability better.

Acknowledgments. This publication is part of the AI4SECIoT project ("Artificial Intelligence for Securing IoT Devices"), funded by the National Cybersecurity Institute (INCIBE), derived from a collaboration agreement signed between the National Institute of Cybersecurity (INCIBE) and the University of Burgos. This initiative is carried out within the framework of the Recovery, Transformation and Resilience Plan funds, financed by the European Union (Next Generation), the project of the Government of Spain that outlines the roadmap for the modernization of the Spanish economy, the recovery of economic growth and job creation, for solid, inclusive and resilient economic reconstruction after the COVID19 crisis, and to respond to the challenges of the next decade.

References

1. Al-Hawawreh, M., Alazab, M., Ferrag, M.A., Hossain, M.S.: Securing the industrial internet of things against ransomware attacks: a comprehensive analysis of the emerging threat landscape and detection mechanisms. J. Netw. Comput. Appl. **223**, 103809 (2024)
2. Casado-Vara, R., Severt, M., Díaz-Longueira, A., Rey, Á.M.D., Calvo-Rolle, J.L.: Dynamic malware mitigation strategies for Iot networks: a mathematical epidemiology approach. Mathematics **12**(2), 250 (2024)
3. Malik, M.S.: IoT malware: a comprehensive survey of threats, vulnerabilities, and mitigation strategies. Int. J. Electr. Crime Invest. **8**(1), 57–66 (2024)
4. Brindha Devi, V., Ranjan, N.M., Sharma, H.: IoT attack detection and mitigation with optimized deep learning techniques. Cybern. Syst. **55**(7), 1702–1728 (2024)
5. Pavica, C., Swanson, G., Whitaker, R., Johansson, S.: A feedback controlled optimization approach to minimize ransomware propagation in internet of things networks (2024)
6. Weston, J., Tolić, D., Palunko, I.: Application of Hamilton–Jacobi–Bellman Equation/Pontryagin's principle for constrained optimal control. J. Optim. Theory Appl. **200**(2), 437–462 (2024)
7. Salehnia, T., et al.: An optimal task scheduling method in IoT-Fog-Cloud network using multi-objective moth-flame algorithm. Multimedia Tools Appl. **83**(12), 34351–34372 (2024)
8. Fleming, W.H., Raymond W.R.: Deterministic and stochastic optimal control. Vol. 1. Springer Science & Business Media (2012)
9. Crandall, M.G., Lions, P.L.: Viscosity solutions of Hamilton-Jacobi equations. Trans. Am. Math. Soc. **277**(1), 1–42 (1983)
10. Gelashvili, K.: The existence of optimal control on the basis of Weierstrass's theorem. J. Math. Sci. **177**, 373–382 (2011)
11. Sniedovich, M.: A new look at Bellman's principle of optimality. J. Optim. Theory Appl. **49**, 161–176 (1986)
12. Hu, S., Qiu, W., Chen, H.: A backward Euler difference scheme for the integro-differential equations with the multi-term kernels. Int. J. Comput. Math. **97**(6), 1254–1267 (2020)
13. Jannelli, A.: A finite difference method on quasi-uniform grids for the fractional boundary-layer Blasius flow. Math. Comput. Simul. **215**, 382–398 (2024)
14. Chang, L., Gong, W., Jin, Z., Sun, G.Q.: Sparse optimal control of pattern formations for an SIR reaction-diffusion epidemic model. SIAM J. Appl. Math. **82**(5), 1764–1790 (2022)
15. Gatto, N.M., Schellhorn, H.: Optimal control of the SIR model in the presence of transmission and treatment uncertainty. Math. Biosci. **333**, 108539 (2021)
16. Fabbri, G.: A viscosity solution approach to the infinite-dimensional HJB equation related to a boundary control problem in a transport equation. SIAM J. Control. Optim. **47**(2), 1022–1052 (2008)

Cybersecurity Taxonomies: Comparative Analysis of Leading IoT Datasets for AI-Driven Security

Virginia Martinez-Fuentes[1](✉), Ángel Arroyo[1], Diego Granados-López[2], and Álvaro Herrero[1]

[1] Grupo de Inteligencia Computacional Aplicada (GICAP), Departamento de Digitalización, Escuela Politécnica Superior, Universidad de Burgos, Avda. Cantabria s/n, 09006 Burgos, Spain
{vmfuentes,aarroyop,dgranados,ahcosio}@ubu.es
[2] SWIFT Solar and Wind Feasibility Technologies, Universidad de Burgos, Avda. Cantabria s/n, 09006 Burgos, Spain

Abstract. This paper outlines emerging security strategies and frameworks, focusing on taxonomies designed to categorize aspects of cybersecurity, particularly IoT security incidents, including threats and attacks. A comparative analysis of the *TON_IoT*, *Edge-IIoTset*, and *CICIoT2023* datasets is presented, examining their classifications of IoT threats and attacks. Additionally, the ENISA Taxonomy adopted by INCIBE-CERT CSIRT is reviewed and compared with the taxonomies of NIST CSF 2.0, MITRE ATT&CK, and OWASP, which incorporates STRIDE. Next, the taxonomies of the *TON_IoT*, *Edge-IIoTset*, and *CICIoT2023* datasets are mapped to the ENISA and INCIBE Taxonomy, correlating the categories of cyber threats and attacks identified in each dataset with these classification systems. Aiming to enhance AI-driven solutions for IoT cybersecurity, this study underscores both the similarities and differences in the number and names of categories of threats and attacks, and in the labels assigned to these incidents across the analyzed taxonomies and datasets.

Keywords: Cybersecurity · IoT · Taxonomy · Machine Learning

1 Introduction

The Internet of Things (IoT) [11] refers to a network of cyber-physical devices [9] that communicate and exchange data with users over the Internet, and are fundamental to Cyber-Physical Systems (CPS) [9]. These interconnected systems enable automation and support applications across sectors including manufacturing, transportation, healthcare, agriculture, education, commerce, military, and consumer applications such as smart homes [20]. Therefore, given its pervasive and heterogeneous nature, the IoT, particularly in smart homes, exemplifies the need for enhanced security due to its vulnerability to cybercriminals [12].

E. Corchado et al. (Eds.): CISIS 2025, CCIS 2807, pp. 157–168, 2026.
https://doi.org/10.1007/978-3-032-19770-2_15

As the number of IoT devices continues to grow, it is expected to surpass 32 billion by 2030 [24]. In response, advancements in data-driven approaches to Intrusion Detection Systems (IDS) [9] have paved the way for Artificial Intelligence (AI) to bolster IoT security. As a result, leveraging Machine Learning (ML) in AI-driven cybersecurity has become pivotal to IoT, aiming to detect and mitigate cyber incidents with a focus on threats and attacks [22].

Acknowledging that the choice of dataset influences the effectiveness of AI techniques for Intrusion Detection Systems (IDS), there is a need for a universal benchmark feature set [21]. Consequently, this paper examines the classification system of IoT security incidents across three leading datasets: *TON_IoT* [3], *Edge-IIoTset* [7], and *CICIoT2023* [15]. *TON_IoT* organizes these incidents into eight primary categories, *Edge-IIoTset* separates them into five, while *CICIoT2023* identifies seven, with each dataset using its own classification system. Likewise, efforts are underway toward the development of a unified taxonomy of cyber incidents, which would provide a reference baseline [5]. Devising such a standardized classification system would facilitate cybersecurity incident management [10], ultimately streamlining the process of classifying IoT threats and attacks when reported to Computer Security Incident Response Teams (CSIRTs) [16].

This paper presents an overview of established cybersecurity taxonomies applicable to the IoT, including those provided by ENISA [6], INCIBE-CERT CSIRT [10], NIST's *Cybersecurity Framework (CSF) 2.0* [17], *Enterprise Matrix* by MITRE ATT&CK [13], and OWASP's *STRIDE*-based approach [18]. It compares these taxonomies and maps the author-defined classification systems of *TON_IoT*, *Edge-IIoTset*, and *CICIoT2023* to the *ENISA Taxonomy* (to which INCIBE has adhered).

Subsequent to the introduction, this document is structured as follows: 2. Selected Datasets, 3. Taxonomy Analysis, 4. Mapping IoT Security Datasets' Attack Classifications to the ENISA and INCIBE Taxonomy, and 5. Conclusion.

2 Dataset Selection

Given the range of cybersecurity datasets available for IoT systems, selecting those most suitable for this study required a structured and reproducible methodology. Relevant literature was retrieved from the Web of Science database using a query that incorporated the following topic search terms: ('IoT' OR 'Internet of Things' OR 'Cyber-Physical Systems') AND ('Taxonomy of Vulnerabilities' OR 'Taxonomy of Attacks' OR 'Dataset' OR 'Datasets') AND ('Artificial Intelligence' OR 'Machine Learning' OR 'Deep Learning' OR 'Supervised Learning' OR 'Unsupervised Learning' OR 'Reinforcement Learning') AND ('Security' OR 'Cybersecurity' OR 'Cyber Security' OR 'Cyber Threats' OR 'Attack Detection' OR 'Security Challenges' OR 'Security Solutions').

This initial query returned 5,394 records during the literature review phase (mid-2025) of this study. To narrow the scope to the most relevant contributions, the results were subsequently refined by applying the following filters:

document type (Article), publication years (2020–2025), citation status (Highly Cited Papers), and language (English). As a result of these refinements, the number of search results was reduced to 92 papers, from which the only three that published and made publicly available IoT security datasets were selected: *TON_IoT* [3], *Edge-IIoTset* [7], and *CICIoT2023* [15].

These datasets have demonstrated a significant academic impact, with a combined total of 984 citations as indexed by the Web of Science at the time this study was conducted. Individually, *TON_IoT* has received 373 citations, *Edge-IIoTset* 369 citations, and *CICIoT2023* 250 citations as of mid-2025. Their consistent citation growth over recent years reflects widespread adoption and sustained relevance within the field of cybersecurity for IoT systems, a trend further reinforced by the credibility of the institutions behind their development. *TON_IoT* was created by the Intelligent Security Group (ISG) at the University of New South Wales (UNSW), Australia. *Edge-IIoTset* was developed collaboratively by researchers from Guelma University and Annaba University (Algeria), De Montfort University (United Kingdom), and Edith Cowan University (Australia). *CICIoT2023* was curated by the Canadian Institute for Cybersecurity at the University of New Brunswick (UNB), Canada.

For this study, one comma-separated values (CSV) file was selected from each dataset for analysis. The following summarizes the cybersecurity taxonomies defined in the original papers and reflected in the selected files.

- **TON_IoT** [3]: The paper classifies cyber incidents into eight groups: backdoor, denial of service (DoS) and distributed DoS (DDoS), injection, man-in-the-middle (MITM), password cracking, ransomware, scanning, and cross-site scripting (XSS). Correspondingly, its *Processed_Network_dataset_1.csv* [3] contains 46 columns, 44 of which are features, along with two additional columns: *label* for binary classification (0 for non-attack and 1 for attack) and *type* for multiclass classification (with 10 possible values corresponding to different attack types, including one for non-attack) [14].
- **Edge-IIoTset** [7]: The paper considers five categories: DoS/DDoS, information gathering, injection, malware, and MITM. Additionally, its *ML-EdgeIIoT-dataset.csv* [7] comprises 63 columns, 61 of which are features, with two additional columns: *Attack_Label* for binary classification (0 for non-attack and 1 for attack) and *Attack_Type* for multiclass classification (15 types, including one for non-attack) [1].
- **CICIoT2023** [15]: The paper features a taxonomy of seven categories: brute force, DDoS, DoS, Mirai, reconnaissance (Recon), spoofing, and web-based threats and attacks. Its *Merged01.csv* [15] contains 40 columns, 39 of which are features, with one additional column (*Label*) for multiclass classification (covering 34 types of attacks, including one category for non-attack) [23].

Considering these combined criteria, the *TON_IoT*, *Edge-IIoTset*, and *CICIoT2023* datasets are particularly suitable for this study, which compares IoT cybersecurity taxonomies to enhance the development of AI-driven solutions for IoT security.

3 Taxonomy Analysis

3.1 Taxonomies for Cyber Incident Classification: The ENISA and INCIBE Approach

The European Union Agency for Cybersecurity (ENISA) has been actively developing a benchmark framework for classifying cyber incidents. Building on the *eCSIRT.net version mkVI Incident Taxonomy* [4], ENISA released the *Reference Incident Classification Taxonomy* report [5] and established the Reference Security Incident Taxonomy Working Group (RSIT WG) [6]. The RSIT WG then published a public GitHub repository containing the *ENISA Reference Incident Classification Taxonomy Task Force* [6]. In the *working_copy/humanv1.md* file within this repository, the taxonomy is presented in tabular format, categorizing incidents into 11 distinct types: 10 core categories and one additional category for testing purposes. Each category is further broken down into real-world examples and detailed descriptions [6].

The National Cybersecurity Institute (INCIBE) CSIRT of Spain, through its Computer Emergency Response Team (CERT), has adopted ENISA's classification system as outlined in its *National Cyber Incident Reporting and Management Guide* [10]. For the purposes of this paper, the taxonomy used by both ENISA and INCIBE-CERT CSIRT will henceforth be referred to as the 'ENISA and INCIBE Taxonomy'.

3.2 Taxonomies for Cyber Risk Management and Governance: The NIST CSF 2.0 Framework

The *Cybersecurity Framework (CSF) 2.0* white paper [17], developed by the United States National Institute of Standards and Technology (NIST), provides recommendations for cybersecurity risk management, supported by a taxonomy focused on security objectives [17]. Although originally designed for U.S.-based organizations, it is also adaptable for use by entities worldwide [19]. This framework is organized into three core components: the CSF Core, CSF Organizational Profiles, and CSF Tiers [17]. The CSF Core serves as the foundation of the NIST CSF 2.0 framework, outlining a hierarchical structure for cybersecurity governance through six core functions: *Govern*, *Identify*, *Protect*, *Detect*, *Respond*, and *Recover*, which can also be applied to the IoT domain [17].

3.3 Taxonomies for Cyber Threat Modeling: The MITRE ATT&CK and OWASP STRIDE Methodologies

Two widely adopted methodologies for cyber threat modeling, extendable to the IoT, are MITRE ATT&CK [13] and the Open Web Application Security Project (OWASP)-STRIDE approach [18].

- **MITRE ATT&CK** [13]: Classifies adversary tactics, techniques, and procedures (TTPs) based on real-world incidents. The *MITRE ATT&CK Enterprise Matrix* [13] consists of 14 adversary tactics, each associated with a range of techniques, totaling 236, as well as additional sub-techniques.

- **OWASP-STRIDE** [18]: Focuses on 6 threat categories: Spoofing, Tampering, Repudiation, Information Disclosure, DoS, and Elevation of Privilege, along with 22 mitigation techniques.

The MITRE ATT&CK knowledge base covers TTPs across various environments, including the IoT [2], and, in addition to its STRIDE-based methodology, OWASP has compiled a unified list of the top IoT security considerations: the *OWASP IoT Top 10* [8].

3.4 Brief Comparative Analysis of Cybersecurity Taxonomies: ENISA and INCIBE, NIST CSF 2.0, MITRE ATT&CK, and OWASP-STRIDE

While NIST CSF 2.0 offers a high-level taxonomy focused on cybersecurity outcomes and practices aimed at governance and risk management [17], the ENISA and INCIBE Taxonomy primarily classifies cyber incidents (specifically threats and attacks) to facilitate incident reporting, management, and response [6]. In contrast, both the MITRE ATT&CK and OWASP-STRIDE methodologies are designed for threat modeling [13,18], each employing distinct approaches. MITRE ATT&CK provides detailed insights into adversaries' tactics, techniques, and procedures during attacks [13], whereas OWASP-STRIDE offers a framework for evaluating potential threats [18].

These taxonomies are extendable to the IoT, with each focusing on different aspects of cybersecurity: the ENISA and INCIBE Taxonomy classifies incidents, NIST CSF 2.0 addresses risks, and MITRE ATT&CK and OWASP-STRIDE provide methodologies for threat modeling. Table 1 presents a comparative analysis of these IoT cybersecurity taxonomies, highlighting their scope, categories, and subcategories.

Table 1. Comparison of Cybersecurity Taxonomies

Taxonomy	Focus	Categories	Subcategories
ENISA [6] and INCIBE [10]	Incident Classification	11 Cyber Incidents	38 Examples and its description
NIST CSF 2.0 [17]	Risk Governance	6 Core Functions	22 Core Categories with multiple subcategories
MITRE ATT&CK [13]	Threat Modeling	14 Adversary Tactics	236 Techniques and various sub-techniques
OWASP STRIDE [18]	Threat Modeling	6 Threat Types	22 Mitigation Techniques

4 Mapping IoT Security Datasets' Attack Classifications to the ENISA and INCIBE Taxonomy

The ENISA and INCIBE Taxonomy consists of 11 high-level incident categories, including one reserved for testing: *Abusive Content, Malicious Code, Information Gathering, Intrusion Attempts, Intrusions, Availability, Information Content Security, Fraud, Vulnerabilities, Other*, and *Test.*

To explore how current IoT security datasets align with this taxonomy, the attack classifications from *TON_IoT*, *Edge-IIoTset*, and *CICIoT2023* are analyzed. Each dataset defines its own set of attack labels and conceptual categories, as introduced by the authors in their respective publications. These labels and groupings were mapped to the ENISA and INCIBE taxonomy based on the conceptual definitions of the attacks, as well as the descriptions and examples provided by ENISA and INCIBE. This mapping serves to visualize similarities and differences in terminology, granularity, and thematic structure across datasets and taxonomy frameworks.

Tables 2, 3, and 4 present the individual mappings for each dataset, providing a label-to-taxonomy alignment while preserving the internal classification logic proposed by the dataset authors.

Table 2. TON_IoT Dataset Labels (10) and Categories (10) Mapped to ENISA and INCIBE Categories

TON_IoT Labels	TON_IoT Categories	ENISA and INCIBE Taxonomy
Backdoor	Backdoor	Intrusion Attempts
DDoS	DDoS	Availability
DoS	DoS	Availability
Injection	Injection	Intrusions
MITM	MITM	Information Gathering
Normal	Normal	Test
Password	Password Cracking	Intrusion Attempts
Ransomware	Ransomware	Information Content Security
Scanning	Scanning	Information Gathering
XSS	XSS	Intrusion Attempts

Table 3. Edge-IIoTset Dataset Labels (15) and Categories (6) Mapped to ENISA and INCIBE Categories

Edge-IIoTset Labels	Edge-IIoTset Categories	ENISA and INCIBE Taxonomy
Backdoor	Malware	Intrusion Attempts
DDoS_HTTP	DoS/DDoS	Availability
DDoS_ICMP	DoS/DDoS	Availability
DDoS_TCP	DoS/DDoS	Availability
DDoS_UDP	DoS/DDoS	Availability
Fingerprinting	Information Gathering	Information Gathering
MITM	Man-in-the-middle: DNS and ARP spoofing	Information Gathering
Normal	Normal	Test
Password	Malware	Intrusion Attempts
Port_Scanning	Information Gathering	Information Gathering
Ransomware	Malware	Information Content Security
SQL_injection	Injection	Intrusions
Uploading	Injection	Malicious Code
Vulnerability_scanner	Information Gathering	Information Gathering
XSS	Injection	Intrusion Attempts

Table 4. CICIoT2023 Dataset Labels (34) and Categories (8) Mapped to ENISA and INCIBE Categories

CICIoT2023 Labels	CICIoT2023 Categories	ENISA and INCIBE Taxonomy
BACKDOOR_MALWARE	Web	Intrusion Attempts
BENIGN	Benign	Test
BROWSERHIJACKING	Web	Malicious Code
COMMANDINJECTION	Web	Intrusions
DDOS-ACK_FRAGMENTATION	DDoS	Availability
DDOS-HTTP_FLOOD	DDoS	Availability
DDOS-ICMP_FLOOD	DDoS	Availability
DDOS-ICMP_FRAGMENTATION	DDoS	Availability
DDOS-PSHACK_FLOOD	DDoS	Availability
DDOS-RSTFINFLOOD	DDoS	Availability
DDOS-SLOWLORIS	DDoS	Availability
DDOS-SYN_FLOOD	DDoS	Availability
DDOS-SYNONYMOUSIP_FLOOD	DDoS	Availability
DDOS-TCP_FLOOD	DDoS	Availability
DDOS-UDP_FLOOD	DDoS	Availability

(*continued*)

Table 4. (*continued*)

CICIoT2023 Labels	CICIoT2023 Categories	ENISA and INCIBE Taxonomy
DDOS-UDP_FRAGMENTATION	DDoS	Availability
DICTIONARYBRUTEFORCE	Brute force	Intrusion Attempts
DNS_SPOOFING	Spoofing	Information Gathering
DOS-HTTP_FLOOD	DoS	Availability
DOS-SYN_FLOOD	DoS	Availability
DOS-TCP_FLOOD	DoS	Availability
DOS-UDP_FLOOD	DoS	Availability
MIRAI-GREETH_FLOOD	Mirai	Availability
MIRAI-GREIP_FLOOD	Mirai	Availability
MIRAI-UDPPLAIN	Mirai	Availability
MITM-ARPSPOOFING	Spoofing	Information Gathering
RECON-HOSTDISCOVERY	Recon	Information Gathering
RECON-OSSCAN	Recon	Information Gathering
RECON-PINGSWEEP	Recon	Information Gathering
RECON-PORTSCAN	Recon	Information Gathering
SQLINJECTION	Web	Intrusions
UPLOADING_ATTACK	Web	Malicious Code
VULNERABILITYSCAN	Recon	Information Gathering
XSS	Web	Intrusion Attempts

To complement these mappings and provide a broader comparative perspective, Table 5 presents a taxonomy-centric view. It organizes attack labels from the three datasets according to the 11 ENISA and INCIBE categories, highlighting overlaps and distinctions, and helping illustrate how cybersecurity taxonomies differ across datasets and in relation to the ENISA and INCIBE framework.

Table 5. Cross-Dataset Comparison of Cyber Attack Labels Mapped to ENISA and INCIBE Incident Taxonomy

TON_IoT Labels	Edge-IIoTset Labels	CICIoT2023 Labels	ENISA and INCIBE Taxonomy
–	–	–	Abusive Content
–	Uploading	BROWSERHIJACKING UPLOADING_ATTACK	Malicious Code
MITM Scanning	Port_Scanning Fingerprinting Vulnerability_scanner MITM	MITM-ARPSPOOFING RECON-HOSTDISCOVERY RECON-OSSCAN RECON-PINGSWEEP RECON-PORTSCAN DNS_SPOOFING VULNERABILITYSCAN	Information Gathering
Backdoor Password XSS	Backdoor Password XSS	DICTIONARYBRUTEFORCE BACKDOOR_MALWARE XSS	Intrusion Attempts
Injection	SQL_injection	SQLINJECTION COMMANDINJECTION	Intrusions
DoS DDoS	DDoS_HTTP DDoS_ICMP DDoS_TCP DDoS_UDP	DDOS-ACK_FRAGMENTATIONv DDOS-HTTP_FLOOD DDOS-ICMP_FLOOD DDOS-ICMP_FRAGMENTATION DDOS-PSHACK_FLOOD DDOS-RSTFINFLOOD DDOS-SLOWLORIS DDOS-SYN_FLOOD DDOS-SYNONYMOUSIP_FLOOD DDOS-TCP_FLOOD DDOS-UDP_FLOOD DDOS-UDP_FRAGMENTATIONv DOS-HTTP_FLOOD DOS-SYN_FLOOD DOS-TCP_FLOOD DOS-UDP_FLOOD MIRAI-GREETH_FLOOD MIRAI-GREIP_FLOOD MIRAI-UDPPLAIN	Availability
Ransomware	Ransomware	–	Information Content Security
–	–	–	Fraud
–	–	–	Vulnerabilities
–	–	–	Other
Normal	Normal	BENIGN	Test

5 Conclusion

This paper analyzed IoT security incident classification systems across three major datasets: *TON_IoT*, *Edge-IIoTset*, and *CICIoT2023*. The comparative analysis highlighted the similarities and differences in the categorization schemes used within these datasets, particularly in the number, naming, and level of detail (granularity) of the categories for IoT cyber threats and attacks.

Additionally, this study outlined established cybersecurity taxonomies, such as those from ENISA, INCIBE-CERT CSIRT, the NIST *CSF 2.0*, the MITRE ATT&CK *Enterprise Matrix*, and the OWASP *STRIDE*-based approach, all of which provide differing yet complementary perspectives on IoT security.

Furthermore, the mapping of the *TON_IoT*, *Edge-IIoTset*, and *CICIoT2023* datasets to the *ENISA Taxonomy* (adopted by INCIBE) revealed areas of both alignment and divergence. While certain categories of cyber incidents were consistent across datasets, differences were observed in the naming conventions, as well as in the number and level of detail of labels used within each dataset and when compared to the ENISA and INCIBE Taxonomy.

Overall, this paper aims to enhance AI-driven solutions for IoT cybersecurity by underscoring both the similarities and differences in the number and names of categories of threats and attacks, along with the labels assigned to these cyber incidents across the analyzed taxonomies and datasets.

Future work may follow current research efforts aimed at developing a standardized feature set for IoT cybersecurity datasets. This effort may be complemented by harmonizing attack labels across datasets, which could further improve comparability and support broader generalization in AI-driven cybersecurity for the IoT.

Acknowledgments. This publication is part of the AI4SECIoT project ('Artificial Intelligence for Securing IoT Devices'), funded by the National Cybersecurity Institute (INCIBE) under grant number C032.23, and derived from a collaboration agreement signed between INCIBE and the University of Burgos. This initiative is carried out within the framework of the Recovery, Transformation, and Resilience Plan funds, financed by the European Union (Next Generation), the project of the Government of Spain that outlines the roadmap for the modernization of the Spanish economy, the recovery of economic growth and job creation, for solid, inclusive, and resilient economic reconstruction after the COVID19 crisis, and to respond to the challenges of the next decade.

Disclosure of Interests. The authors declare that this publication is part of the AI4SECIoT project, funded by INCIBE through a collaboration agreement with the University of Burgos, and the corresponding author is enrolled in the Master's program in Cybersecurity Research at the University of León, which maintains internship agreements with INCIBE and includes faculty affiliated with INCIBE. The authors declare that they have no other competing interests relevant to the content of this article.

References

1. Al Nuaimi, T., et al.: A comparative evaluation of intrusion detection systems on the Edge-IIoT-2022 dataset. Intell. Syst. Appl. **20**, 200298 (2023). https://doi.org/10.1016/j.iswa.2023.200298
2. Al-Sada, B., Sadighian, A., Oligeri, G.: Analysis and characterization of cyber threats leveraging the MITRE ATT&CK database. IEEE Access **12**, 1217–1234 (2024). https://doi.org/10.1109/ACCESS.2023.3344680

3. Alsaedi, A., Moustafa, N., Tari, Z., Mahmood, A., Anwar, A.: TON_IoT telemetry dataset: a new generation dataset of iot and iiot for data-driven intrusion detection systems. IEEE Access **8**, 165130–165150 (2020). https://doi.org/10.1109/ACCESS.2020.3022862
4. eCSIRT.net, Arvidsson, J., Stikvoort, D.: Incident Classification / Incident Taxonomy (2015). https://static.www.switch.ch/sites/default/files/2023-11/Incident-Classification-Taxonomy.pdf, Accessed 10 Jul 2025
5. ENISA: Reference Incident Classification Taxonomy (2018). https://www.enisa.europa.eu/publications/reference-incident-classification-taxonomy, Accessed 10 Jul 2025
6. ENISA: Reference Security Incident Classification Taxonomy Tax Force. https://github.com/enisaeu/Reference-Security-Incident-Taxonomy-Task-Force (2018), Accessed 10 Jul 2025
7. Ferrag, M.A., Friha, O., Hamouda, D., Maglaras, L., Janicke, H.: Edge-IIoTset: a new comprehensive realistic cyber security dataset of IoT and IIoT applications for centralized and federated learning. IEEE Access **10**, 40281–40306 (2022). https://doi.org/10.1109/ACCESS.2022.3165809
8. Ferrara, P., Mandal, A.K., Cortesi, A., Spoto, F.: Static analysis for discovering IoT vulnerabilities. Int. J. Softw. Tools Technol. Transfer **23**, 71–88 (2021). https://doi.org/10.1007/s10009-020-00592-x
9. IEEE: 2413–2019 - IEEE Standard for an Architectural Framework for the Internet of Things (IoT) (2020). https://doi.org/10.1109/IEEESTD.2020.9032420
10. INCIBE: National Cyber Incident Reporting and Management Guide. https://www.incibe.es/incibe-cert/guias-y-estudios/guias/guia-nacional-de-notificacion-y-gestion-de-ciberincidentes (2020), Accessed 10 Jul 2025
11. ISO/IEC: ISO/IEC 30141, Internet of Things (IoT) – Reference Architecture, Edition 2.0 (2024). https://www.iso.org/standard/88800.html
12. Iturbe Araya, J.I., Rifà-Pous, H.: Anomaly-based cyberattacks detection for smart homes: a systematic literature review. Internet Things **22**, 100792 (2023) https://doi.org/10.1016/j.iot.2023.100792
13. MITRE ATT&CK: MITRE ATT&CK Knowledge Base of Adversary Tactics and Techniques. https://attack.mitre.org, Accessed 10 Jul 2025
14. Moustafa, N.: A new distributed architecture for evaluating AI-based security systems at the edge: network TON_IoT datasets. Sustain. Cities Soc. **72**, 102994 (2021) https://doi.org/10.1016/j.scs.2021.102994
15. Neto, E.C.P., Dadkhah, S., Ferreira, R., Zohourian, A., Lu, R., Ghorbani, A.A.: CICIoT2023: a real-time dataset and benchmark for large-scale attacks in IoT environment. Sensors **23**(13), 5941 (2023). https://doi.org/10.3390/s23135941
16. NIST: Computer Security Incident Response Team (CSIRT). https://csrc.nist.gov/glossary/term/computer_security_incident_response_team, Accessed 10 Jul 2025
17. NIST: The NIST Cybersecurity Framework (CSF) 2.0. NIST Cybersecurity White Paper (CSWP) NIST CSWP 29 (2024). https://doi.org/10.6028/NIST.CSWP.29
18. OWASP: Threat Modeling Process. https://owasp.org/www-community/Threat_Modeling_Process, Accessed 10 Jul 2025
19. Parmar, M., Miles, A.: Cyber Security Frameworks (CSFs): An assessment between the NIST CSF v2.0 and EU standards. In: 2024 Security for Space Systems (3S), pp. 1–7 (2024) https://doi.org/10.23919/3S60530.2024.10592293
20. Saba, T., Rehman, A., Sadad, T., Kolivand, H., Bahaj, S.A.: Anomaly-based intrusion detection system for IoT networks through deep learning model. Com-

put. Electr. Eng. **99**, 107810 (2022). https://doi.org/10.1016/j.compeleceng.2022.107810

21. Sarhan, M., Layeghy, S., Moustafa, N., Gallagher, M., Portmann, M.: Feature extraction for machine learning-based intrusion detection in IoT networks. Digital Commun. Netw. **10**(1), 205–216 (2024). https://doi.org/10.1016/j.dcan.2022.08.012
22. Sarker, I.H., Khan, A.I., Abushark, Y.B., Alsolami, F.: Internet of Things (IoT) security intelligence: a comprehensive overview, machine learning solutions and research directions. Mob. Netw. Appl. **28**, 296–312 (2023). https://doi.org/10.1007/s11036-022-01937-3
23. Tseng, S.M., Wang, Y.Q., Wang, Y.C.: Multi-class intrusion detection based on transformer for iot networks using CIC-IoT-2023 dataset. Future Internet **16**(8), 284 (2024). https://doi.org/10.3390/fi16080284
24. Vailshery, L.S.: Number of Internet of Things (IoT) connections worldwide from 2022 to 2023, with forecasts from 2024 to 2033 (in billions) (2024). https://www.statista.com/statistics/1183457/iot-connected-devices-worldwide, Accessed 10 Jul 2025

Privacy-Enhancing Federated Learning-Based IDS for IoT Networks Using Post-Quantum Secure Channels and Verifiable Secret Sharing

Lamine Syne[1(✉)], Candelaria Hernández-Goya[2], and Pino Caballero-Gil[2]

[1] Fundación General Universidad de La Laguna, Santa Cruz de Tenerife, Spain
lsyne@fg.ull.es

[2] Universidad de La Laguna, Santa Cruz de Tenerife, Spain
{mchgoya,pcaballe}@ull.edu.es

Abstract. Federated Learning (FL)-based Intrusion Detection Systems (IDS) are emerging as a promising approach for securing IoT networks and preserving data confidentiality. Moreover, FL introduces novel vulnerabilities. These include inference attacks by malicious aggregators, who can extract sensitive information from model updates, as well as malicious clients capable of submitting falsified updates to the aggregator server. Additionally, with the rapid development of quantum computers, existing privacy protection schemes mainly based on Secure Multi-Party Computation (SMPC) will no longer be able to guarantee the data.This paper presents a Secure Aggregation (SA) method that combines Post-Quantum-secure channels for client key exchange and Verifiable Secret Sharing (VSS), achieving resilience against malicious clients. Experiments conducted in an FL-based IDS for real-world IoT networks demonstrate the viability of our proposal.

Keywords: Federated Learning · IoT · Secure Aggregation · Post-quantum Security

1 Introduction

The Internet of Things (IoT) has seen a substantial increase in adoption in recent years. It links a wide range of connected devices, including monitoring sensors, wearable health devices and smart home appliances. However, this proliferation of connected devices has led to a significant increase in security and privacy risks. Federated Learning (FL) based Intrusion Detection Systems (IDS) are essential tools for safeguarding IoT networks. They continuously monitor network traffic and system behavior to identify any sign of malicious activity or anomalies while preserving privacy [1]. Traditionally, IDS solutions are deployed in a centralized manner. In these systems, data collected from all connected devices is analyzed on a central server. While this centralized approach is effective in some respects,

E. Corchado et al. (Eds.): CISIS 2025, CCIS 2807, pp. 169–179, 2026.
https://doi.org/10.1007/978-3-032-19770-2_16

it presents critical issues. In particular, it raises privacy concerns about sensitive data transmitted from individual devices. FL [2] enables the training machine learning models directly on distributed IoT devices. This allows data to remain local while minimizing the need to transfer raw information to a central server. By doing so, this method enhances data privacy and security. It leverages the collective intelligence of multiple devices without exposing sensitive information. Nonetheless, the adoption of FL also introduces a new set of challenges [3], one of which is securing the training process against adversarial threats. Existing privacy-preserving FL methods face the challenge of high communication and computational overhead. Furthermore, the rapid development of quantum computers means that these approaches, mostly based on SMPC, will no longer be able to guarantee data security for participants in the post-quantum era.

This work focuses on exploring the integration of FL techniques to strengthen IoT network security while improving the confidentiality of sensitive data, thereby paving the way for more resilient and privacy-aware IDS.

2 Secure Multi-Party Computation (SMPC) and Secret Sharing

Secure Multi-Party Computation (SMPC) is a cryptographic mechanism that allows parties to jointly compute private inputs without revealing anything but the final result. It addresses the following problem, involving m parties or devices denoted as $P_1, ..., P_m$. Each device P_i, where $i = 1, 2, ..., m$, holds a secret input x_i, for $i = 1, 2, ..., m$, and all devices agree on some function f that takes m inputs. Their goal is to compute $y = f(x_1.....x_m)$ while making sure that the correct value of y is calculated [4] without releasing any information about x_i.

2.1 Shamir's Secret Sharing

Shamir's Secret Sharing (SSS) [5] is a method for splitting a secret into shares where only a certain number of shares (threshold) are needed to recover the original secret. It's called a (t,n)-threshold-based scheme that securely distributes a secret S among n participants such that only t or more participants can reconstruct it. The secret is represented as the constant term of a random polynomial $f(x)$ of degree $t-1$ over a finite field $\mathbb{F}_q$, where q is a prime number greater than S. The polynomial is defined as:

$$f(x) = a_0 + a_1x + a_2x^2 + \cdots + a_{t-1}x^{t-1}, \quad \text{with } a_0 = S.$$

Each participant i receives a share $(x_i, f(x_i))$, where $x_i \neq 0$. To reconstruct the secret, any group of at least t participants can use Lagrange interpolation to compute: $S = f(0) = \sum_{j=1}^{t} f(x_j) \prod_{\substack{1 \leq m \leq t \\ m \neq j}} \frac{x_m}{x_m - x_j}$

This method provides perfect secrecy: fewer than t shares reveal no information about S.

2.2 Verifiable Secret Sharing

Verifiable Secret Sharing (VSS) [6] adds a verifiable feature to the secret sharing mechanism, allowing participants to check whether the retrieved information corresponds to the original secret. A common VSS scheme builds upon Shamir's Secret Sharing and uses public commitments to the polynomial coefficients to ensure share consistency. The dealer selects a random polynomial: $f(x) = a_0 + a_1x + a_2x^2 + \cdots + a_{t-1}x^{t-1}, \quad \text{where } a_0 = S$

To enable verification, the dealer publishes commitments $C_j = g^{a_j} \in \mathbb{G}$, where g is a generator of a cyclic group $\mathbb{G}$ of prime order q. Each participant i receives a share $s_i = f(x_i)$. Participants can verify their share by checking: $g^{s_i} \stackrel{?}{=} \prod_{j=0}^{t-1} C_j^{x_i^j}$

If the equation holds, the participant is convinced that its share is consistent with the committed polynomial. This approach provides both secrecy and verifiability under standard cryptographic assumptions.

3 Related Work

FL has gained major interest in recent years, with numerous applications and use cases, particularly in the field of IoT. In this context, some recent works have proposed its use to improve IDS. In [7] a self-learning system for detecting compromised devices in IoT networks based on FL is introduced. Its findings demonstrate an average detection rate of 95.6% for attacks in just 257 milliseconds, all without triggering any false alarms during evaluation in a real-world deployment. Despite its benefits, FL suffers from different vulnerabilities as described in [1]. To overcome these, Secure Aggregation (SA) is widely deployed in FL. SA is a protocol that ensures the aggregator can only learn the average of the updates of clients while keeping their updates private. The concept was first proposed by Bonawitz et al. [8]. Their protocol is based on a Secure Multi-Party Computation (SMPC) approach which refers to a mechanism that allows parties to jointly compute private inputs without revealing anything but the final result. SAFELearn [9] proposed a generic system to secure aggregation in FL. The framework can be instantiated with one (FHE) or multiple non-colluding servers (SMPC). A secure two-party computation instantiation of SAFELearn applied to a personalized version of [7] achieves convergence with a learning rate of 0.1 while keeping a good trade-off between security and efficiency. The work [10] presents VerifyNet, the first privacy-preserving and verifiable federated learning framework. A double-masking protocol is used to guarantee the confidentiality of users' local gradients during the federated learning. Then, the cloud server is required to provide a proof of the correctness of its aggregated results to each user. Most of the SA protocols use cryptographic primitives such as pseudo-random generators, key agreement protocols, authenticated encryption, or Shamir's secret sharing to guarantee the security of communications between the server and clients, as well as among the clients themselves. However, these primitives are mainly based on traditional cryptography, such as the

Diffie-Hellman key agreement protocol and general-purpose ciphers. As a result, these approaches lack resistance to quantum attacks.

Numerous studies have looked into the integration of post-quantum cryptography (PQC) into federated learning to mitigate quantum computing threats. The work [11] proposes a lattice-based FL protocol where compressed and encrypted model updates are used to reduce communication overhead; however, it does not support dropout recovery or verifiable aggregation. In [12], PQC-based aggregation protocols relying on peer-to-peer share exchange or outdated primitives such as NewHope are proposed, neither supports dropout resilience and both incur high communication overhead. Saidi et al. [13] propose to combine Ciphertext-Policy Attribute-Based Encryption (CP-ABE) to secure models and the Cheon-Kim-Kim-Song (CKKS) homomorphic encryption algorithm to protect model parameters but with substantial latency unsuitable for edge devices. By comparison, our approach leverages the NIST-recommended ML-KEM, Shamir's Secret Sharing, and Verifiable Secret Sharing to achieve efficient, verifiable, and dropout-resilient aggregation with low overhead, making it suitable for deployment in IoT and other resource-constrained environments.

4 Adversary Model

In this work, the adversary is considered a semi-honest aggregator, that is, it follows the protocol honestly but tries to infer sensitive information about the local device data from their model updates during the learning process. In the standard FL setting, the aggregator has access to all local model updates W_i, and clients exchange model updates with the aggregator server, such that malicious aggregators can perform model inference attacks that aim to extract information about training data from the global model updates [14]. However, these attacks acquire minimal aggregated information about the data. Therefore, our objective is to conceal local models from the aggregator, thwarting potent inference attacks while still facilitating the accurate performance of FL-based IDS. Malicious clients are also capable of submitting falsified updates to the aggregator server and emerging quantum computing capabilities introduce additional threats.

5 VSS-Based Secure Aggregation Protocol

In order to preserve the confidentiality of local contributions in a FL context, we propose a SA protocol combining private and pairwise masks between clients, Shamir Secret Sharing (SSS) for client dropout tolerance and a verifiable sharing scheme (VSS) to guarantee the integrity of aggregated data. The protocol contains three key steps described by Algorithm 1, 2 and 3.

In the initialization phase (Algorithm 1) a set of clients (users) $\mathcal{U} = \{U_1, U_2, \ldots, U_n\}$ is considered. Each client U_i generates a private mask b_i and pairwise masks $s_{i,j}$ with all the other clients. The private mask b_i is then shared with the

Algorithm 1. Initialization()

Key Agreement Phase:
ML-KEM: Post Quantum Key Exchange between clients.
Initialization Phase:
for each client $U_i \in \mathcal{U}$ **do**
Generate private mask b_i.
Generate pairwise masks with other clients U_j:

$$s_{i,j} \leftarrow \text{RandomMaskGeneration}()$$

Compute final masking key y_i:

$$y_i = b_i + \sum_{j:i<j} s_{i,j} - \sum_{j:i>j} s_{j,i}$$

Secret-share private mask b_i using SSS:

$$\text{Shares}_{b_i} = \text{SSS.Share}(b_i, t, n)$$

Distribute shares of b_i to clients.
end for

Algorithm 2. LocalTraining (Model: M, Parameters : θ, Private data : d_i)

Local Training Phase:
for each client $U_i \in \mathcal{U}$ **do**
Compute local model weights:

$$W_i = \text{LocalTraining}(M, \theta, \text{d}_i)$$

Mask the model weights:

$$W_{imasked} = W_i + y_i$$

Compute VSS commitments for masked model weights:

$$\text{C}_{W_{imasked}} = \text{VSS.Commit}(W_{imasked})$$

Distribute secret shares of the masked model weights to the aggregator$\mathcal{S}$.
end for

other clients using the secret sharing scheme (SSS) with a threshold t. A previous post-quantum key negotiation (ML-KEM) stage is used to secure channels between clients. In the second step (Algorithm 2), each client trains a local model W_i using its private data and initial model parameters θ. The model is then masked by y_i to give $W_{imasked} = W_i + y_i$ and then cryptographic com-

Algorithm 3. VerifiedAggregation $(W_{i_{masked}}, C_{W_{i_{masked}}})$

Verification Phase:
for aggregator $\mathcal{S}$ **do**
  for each received model weights share $W_{i_{masked}}$ **do**
    Verify model weights share using VSS:

$$\text{Valid} \leftarrow \text{VSS.Verify}(W_{i_{masked}}, C_{W_{i_{masked}}})$$

    if verification fails **then**
      Reject masked model weights from client U_i
    end if
  end for
end for
Aggregation Phase:
Aggregator sums verified masked model weights: $W_{masked} = \sum_{i=1}^{n} W_{i_{masked}}$
If dropout is detected, aggregator reconstructs dropped clients' private masks via SSS: $b_i \leftarrow \text{SSS.Recons}(\text{Shares}_{b_i})$
Aggregator removes all masks using reconstructed private masks and known pairwise masks to recover true aggregate model weights. $W = W_{masked} - \sum_{i \in \text{dropped}} b_i$
Aggregator computes the global model: $\theta \leftarrow \frac{1}{|\mathcal{U}|} W$

mitments $C_{W_{i_{masked}}} = VSS.Commit(W_{i_{masked}})$ are calculated on the masked weights to allow future verifications.

In the final stage, the aggregator server $\mathcal{S}$ receives the masked models and verifies their integrity using commitments (VSS.Verify) and then the valid models are summed to form the masked aggregate W_{masked} (Algorithm 3). If some clients are missing, their b_i private masks are reconstructed using the received SSS shares. The server eliminates all masks (private + pairwise) to recover the real sum of the models. The global model is finally updated by a standard average.

In the proposed protocol, the client performs fundamental tasks related to Algorithm 1, such as local model training and commitment computation as defined in Algorithm 2. In terms of computation, the client establishes shared keys with $n-1$ other clients. The cost of this operation mainly depends on the ML-KEM scheme used and is estimated as $O(n \cdot \text{ML-KEM}_{\text{cost}})$. The generation of private masks b_i, which are random vectors of dimension m, has a cost of $O(m)$. The generation of pairwise masks s_{ij}, their computation with $n-1$ clients, and the secret sharing of the private mask b_i into n shares are collectively estimated at $O(n \cdot m)$, with the dominant cost attributed to the underlying Secret Sharing Scheme (SSS). The local model training, its masking W_i, and the computation of Verifiable Secret Sharing (VSS) commitments for future server-side verification have respective costs of $O(\text{local\_data\_size} \cdot m)$, $O(m)$, and $O(m \cdot \text{VSS}_{\text{cost}})$.

In summary, the total computational cost corresponds to the following expression: $O(n \cdot m + \text{ML-KEM}_{\text{cost}} + m \cdot \text{VSS}_{\text{cost}})$, which can be approximated, after harmonizing notation and omitting lower-order terms, as $O(n + n \cdot m)$.

Regarding communication overhead, a similar analysis was conducted, focusing on the volume of data exchanged among the different entities. The resulting communication cost is estimated as $O(n \cdot m)$.

6 Dataset and Classification Scenarios

To achieve FL-based intrusion detection system for IoT networks FederatedAveraging [2] approach has been experimented using the dataset from the Canadian Institute for Cybersecurity at the University of New Brunswick [15].The study seeks to present a rich IoT attack corpus that supports both the classification of IoT devices and the detection of abnormal behavior, thereby enhancing security-analytics solutions for practical IoT deployments. To compile the dataset, researchers executed 33 separate attack scenarios across a network of 105 IoT devices hosted at the Institute. Seven types of attacks were run on the experiments: distributed denial of service (DDoS), denial of service (DoS), reconnaissance, web-based, brute-force, spoofing, and the Mirai botnet.

To evaluate the performance of the proposed IDS, three different classifications were considered.

1. Binary classification: This scenario distinguishes between normal behavior and attack. It is particularly important when the main goal of the IDS is to detect malicious activity
2. Eight-category classification: Here the system differentiates among the main general categories of attacks, providing more information about the characteristics of the anomalies detected. Seven main attack types and normal behavior are considered
3. Thirty-four-category classification: In this setting, the system is able to detect and classify specific variants of attacks.

To specifically target Mirai-type attacks, we performed a balanced reduction of the dataset to 1% of the original size, followed by aggregation to eight classes and, finally, reduction to two classes (Mirai attacks versus normal behavior). These classification scenarios enable a comprehensive evaluation of the system's performance across varying levels of complexity. This dataset was cleaned, features were selected, normalized, encoded, and checked for possible data leakage.

7 Experimental Evaluation and Results

This section presents the experimental evaluation of our privacy-enhancing methods on a FL-based IDS using a dataset collected from real-world experiments with IoT device. Specifically, we focus on detecting Mirai attacks through binary classification as Mirai is notorious malware that targets vulnerable IoT devices, such as routers, IP cameras and other connected gadgets. We analyze the tradeoff between model efficiency and the overhead of Secure Aggregation by leveraging various metrics.

Table 1 shows how long it takes for the ML-KEM-based key exchange to run at different security levels. Even for the highest level of security (ML-KEM-1024), the key generation and exchange times remain under one millisecond. These low runtimes demonstrate that Post-Quantum secure channels can be integrated into IoT devices without causing significant delays. This is particularly relevant in environments where computational resources and real-time performance are critical.

Table 1. ML-KEM based key exchange runtime (ms)

Algorithm	Key Gen. (ms)	Encap. (ms)	Decap. (ms)
ML-KEM-512	0.2490	0.4535	0.3481
ML-KEM-768	0.4073	0.5127	0.4004
ML-KEM-1024	0.4658	0.5130	0.5931

For the FL framework, the FederatedAveraging [2] aggregation algorithm is adopted and evaluated with three models: Logistic Regression (LR), Long Short-Term Memory (LSTM), and Multi-Layer Perceptron (MLP). Table 2 presents the accuracy, precision and recall metrics. We implement the VSS-based secure aggregation protocol, which includes a post-quantum key exchange based on ML-KEM. The experiments are performed on a Raspberry Pi 4 model B platform with **CPU** 64-bit quad-core Cortex-A72 and **Memory** 8 GB LPDDR4 RAM.

Table 2. Model Results

Model	*Accuracy%*	*Precision%*	*Recall%*
Logistic Regression	99.37	99.37	99.37
LSTM	99.44	99.96	99.25
MLP	99.91	99.98	99.89

In Fig. 2, it is illustrated how well each model performs when combined with the Secure Aggregation protocol. The MLP model, in particular, reaches almost 98,9% accuracy, which is impressive considering the added privacy layer. Even the LSTM and Logistic Regression models maintain solid results. These outcomes indicate that the integration of verifiable secret sharing and post-quantum secure channels does not compromise detection performance.

Table 3. VSS-based Secure Aggregation overhead (n: number of client in the FL training, m: dimension of local model weight vector)

Overhead	Client side	Server side
Communication	$\mathcal{O}(n \cdot m)$	$\mathcal{O}(n^2 \cdot m)$
Computation	$\mathcal{O}(n + n \cdot m)$	$\mathcal{O}(n^2 + n \cdot m)$

Table 3 summarizes the approximate computation and communication overhead of the proposed SA protocol (described in Algorithms 1, 2, and 3). Figure 1 compares this overhead to other SMPC-based Secure Aggregation protocols [8,16,17]. Our proposal is generally less expensive in computation in both side client and server than the considered well-established SA protocol in Federated Learning [8]. Regarding communication, our protocol shows higher costs because of the commitment generation process and verification, enabling the server to verify future model updates. Figure 2 shows a simulation of the SA protocol that involves injecting noise related to masking and share reconstruction. It is shown that the accuracy of the models during the FL training rounds is not significantly impacted, with only a slight difference observed.

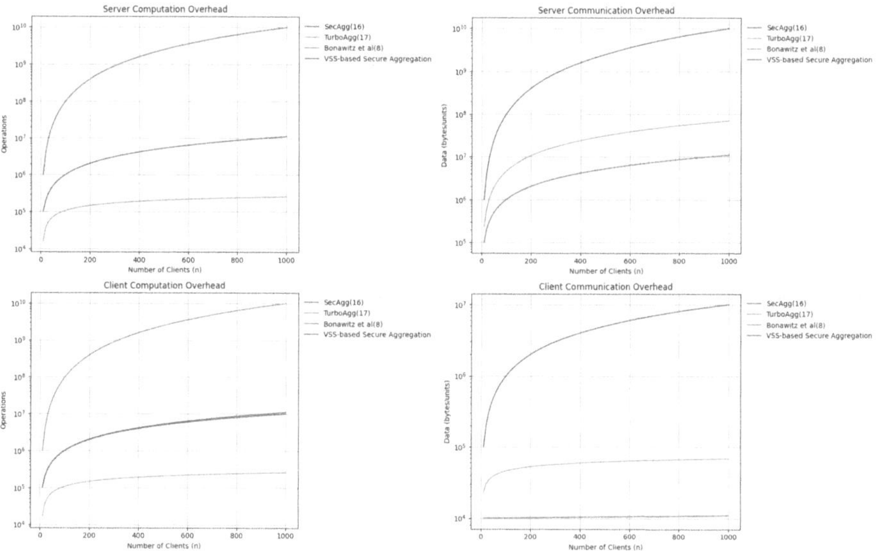

Fig. 1. VSS-based Secure Aggregation overhead comparison.

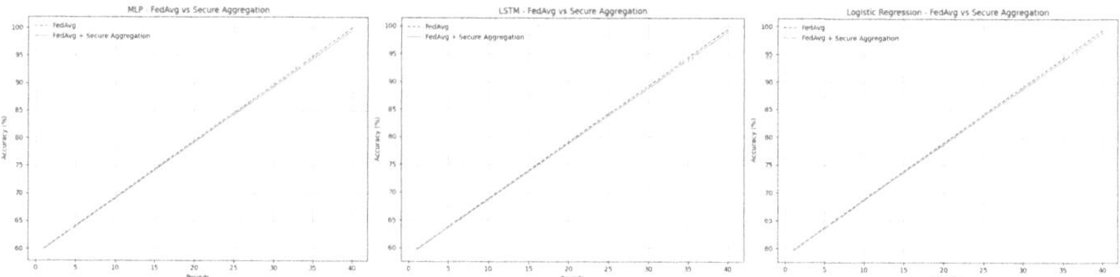

Fig. 2. FedAverage FL with VSS based-Secure Aggregation.

In summary, these experiments confirm that our approach can effectively balance privacy, security, and detection performance in real-world IoT scenarios. By combining verifiable secret sharing and post-quantum key exchange, we've managed to build a FL-based IDS that's both resilient against quantum threats and feasible for devices with limited resources. We believe these results are a strong step toward bringing truly secure FL-based intrusion detection systems into the IoT world.

8 Conclusion and Future Work

In this paper, a Federated Learning-based Intrusion Detection System (IDS) has been built using a dataset collected from real-world experiments with IoT devices. For Federated Learning, FederatedAveraging as the aggregation algorithm has been used, with three different models, namely Logistic Regression, LSTM, and MLP, evaluated. To improve privacy as well as protect against malicious clients, a VSS-based Secure Aggregation protocol has been designed, which runs over a Post-Quantum secure channel. Simulation demonstrates this protocol adding negligible overhead along with lower costs than other SMPC-based methods. Secure Aggregation integration has not hindered model performance considerably, with the MLP model attaining a maximum accuracy of 98.9%.

Future research will extend the current IDS in order to detect other types of attacks. We also intend to compare other quantum-resistance Secure Aggregation protocols, including ones using approximate homomorphic encryption (e.g., methods using CKKS). The Verifiable Secret Sharing implemented here is secured under the hardness of the discrete logarithm problem, but its client-dropping resilience—achieved via Shamir's Secret Sharing—remains to be thoroughly analyzed. We plan to examine this client-dropping tolerance more precisely. We intend to incorporate lattice-based methods in order to improve the verification process against quantum attacks.

Acknowledgments. This work was supported by the Cybersecurity Chair of the University of La Laguna (C065/23) and the strategic project SCITALA (C064/23), both funded by INCIBE through the Recovery, Transformation, and Resilience Plan (Next Generation EU), financed by the European Union; the project PID2022-138933OB-I00 "ATQUE," funded by MCIN/AEI/10.13039/501100011033 and co-financed by the European Regional Development Fund (ERDF, EU); and the 2023DIG28 project, funded by CajaCanarias and the "la Caixa" Foundation.

References

1. Santos, L., Rabadao, C., Gonçalves, R.: Intrusion detection systems in internet of things: a literature review. In 2018 13th Iberian Conference on Information Systems and Technologies (CISTI), pp. 1–7. IEEE (2018)
2. McMahan, H.B., Moore, E., Ramage, D., Agüera y Arcas, B.: Federated learning of deep networks using model averaging. arXiv preprint arXiv:1602.05629, vol. 2, no. 2, pp. 15–18 (2016)

3. Almutairi, S., Barnawi, A.: Federated learning vulnerabilities, threats and defenses: a systematic review and future directions. Internet of Things **24**, 100947 (2023)
4. Cramer, R., Damgård, I.B., Nielsen, J.B.: Secure Multiparty Computation and Secret Sharing. Cambridge University Press, Cambridge (2015)
5. Shamir, A.: How to share a secret. Commun. ACM **22**(11), 612–613 (1979)
6. Shyamalendu Kandar and Bibhas Chandra Dhara: A verifiable secret sharing scheme with combiner verification and cheater identification. J. Inf. Secur. Appl. **51**, 102430 (2020)
7. Nguyen, T.D., Marchal, S., Miettinen, M., Fereidooni, H., Asokan, N., adeghi, A.R.: Dot: a federated self-learning anomaly detection system for IoT. In: IEEE 39th International Conference on Distributed Computing Systems (ICDCS), pp. 756–767. IEEE Computer Society. Los Alamitos (2019)
8. Bonawitz, K., et al.: Practical secure aggregation for federated learning on user-held data. arxiv 2016. arXiv preprint arXiv:1611.04482, 13
9. Fereidooni, H., et al.: Safelearn: secure aggregation for private federated learning. In: IEEE Security and Privacy Workshops (SPW), pp. 56–62. IEEE Computer Society, Los Alamitos (2021)
10. Guowen, X., Li, H., Liu, S., Yang, K., Lin, X.: Verifynet: secure and verifiable federated learning. IEEE Trans. Inf. Forensics Secur. **15**, 911–926 (2019)
11. Xu, P., Hu, M., Chen, T., Wang, W., Jin, H.: LaF: lattice-based and communication-efficient federated learning, vol. 17, pp. 2483–2496
12. Zhang, X., Deng, H., Wu, R., Ren, J., Ren, Y.: PQSF: post-quantum secure privacy-preserving federated learning, vol. 14, no. 1, p. 23553. Nature Publishing Group (2024)
13. Saidi, A., Amira, A., Nouali, O.: Securing decentralized federated learning: cryptographic mechanisms for privacy and trust. Clust. Comput. **28**(2), 1–17 (2025)
14. Bouacida, N., Mohapatra, P.: Vulnerabilities in federated learning. IEEE Access **9**, 63229–63249 (2021)
15. Neto, E.C.P., Dadkhah, S., Ferreira, R., Zohourian, A., Lu, R., Ghorbani, A.A.: CICIoT2023: a real-time dataset and benchmark for large-scale attacks in IoT environment. Sensors **23**(13), 5941 (2023)
16. Li, K.H., de Gusmão, P.P.B., Beutel, D.J., Lane, N.D.: Secure aggregation for federated learning in flower. In Proceedings of the 2nd ACM International Workshop on Distributed Machine Learning, pp. 8–14 (2021)
17. So, J., Güler, B., Avestimehr, A.S.: Turbo-aggregate: breaking the quadratic aggregation barrier in secure federated learning. IEEE J. Sel. Areas Inf. Theory **2**(1), 479–489 (2021)

Federated Learning for the Detection of Attacks on IoT Environment

Alejandro Diez(✉), Jaime Rincon, and Daniel Urda

Grupo de Inteligencia Computacional Aplicada (GICAP), Departamento de Digitalización, Escuela Politécnica Superior, Universidad de Burgos, Av. Cantabria s/n, 09006 Burgos, Spain
{adbermejo,jarincon,durda}@ubu.es

Abstract. The rapid expansion of Internet of Things (IoT) devices has significantly increased the attack surface, necessitating the development of robust and privacy-preserving intrusion detection systems. This research delves into the application of Federated Learning (FL) to detect network-based attacks in IoT environments, utilizing the NF-ToN-IoT dataset. We compare a centralized machine learning model with a federated counterpart, employing the Federated Averaging (FedAvg) algorithm. Both models employ a Multi-Layer Perceptron (MLP) architecture trained on NetFlow features for binary classification of benign versus malicious traffic. Our experimental results reveal that the federated model achieves comparable and even slightly superior performance across various metrics, including precision, recall, and F1-score, while preserving data privacy by decentralizing the training data. These findings underscore the potential of FL as a viable alternative to traditional intrusion detection systems in real-world, privacy-sensitive IoT scenarios.

Keywords: intrusion detection · cybersecurity · iot · federated learning · machine learning · supervised learning

1 Introduction and Previous Work

In recent years, the exponential increase in Internet-connected devices has given rise to a highly complex and widely distributed Internet of Things (IoT) ecosystem. Spanning from smart homes to industrial infrastructure, projections estimate that by 2030, approximately 40 billion IoT devices will be in operation worldwide [7]. This massive expansion has not only increased data generation but has also broadened the attack surface, making IoT devices particularly vulnerable to security threats. According to Forescout's Riskiest Connected Devices of 2025 report, routers account for more than 50% of devices with critical vulnerabilities, surpassing other devices such as computers and wireless access points [8]. This finding highlights the increasing exposure of IoT devices to cyberattacks, particularly those that are part of the network infrastructure.

E. Corchado et al. (Eds.): CISIS 2025, CCIS 2807, pp. 180–189, 2026.
https://doi.org/10.1007/978-3-032-19770-2_17

To mitigate these threats, Artificial Intelligence (AI), and particularly Machine Learning (ML), has emerged as a powerful tool for the proactive detection of threats. ML models can analyze large volumes of network traffic data to identify anomalous patterns that may indicate malicious activity, even before they materialize into actual attacks. However, traditional ML methods often rely on centralized data storage and processing on remote servers. This approach raises critical challenges such as privacy risks, high bandwidth demands, and latency issues, especially in resource-limited settings like IoT networks.

To address these limitations, Federated Learning (FL) offers a decentralized ML approach that enables model training across distributed devices without sharing raw data. This paradigm ensures that sensitive information remains local while allowing for collaborative training. Despite its benefits, applying FL in IoT contexts presents unique challenges, including device heterogeneity, communication efficiency, and robustness against compromised nodes [3].

In this work, we aim to evaluate the feasibility of using FL for anomaly detection in IoT networks, in comparison with traditional centralized ML. To this end, we propose a simulated scenario involving a network with two devices acting as clients, each responsible for processing data and training the same model in a distributed manner. This is achieved by splitting the NF-ToN-IoT dataset [15], a recent and comprehensive intrusion detection dataset tailored to IoT and network-based attacks, originally used to train the model in a traditional centralized framework. To assess the effectiveness of this method, we compare the performance of the federated training approach with that of the same model trained in a centralized setting on a single machine. Our experimental design represents an initial step towards evaluating more complex scenarios, ultimately aiming to apply the proposed methodology in a real-world environment.

The remainder of this paper is organized as follows. Section 2 introduces the NF-ToN-IoT dataset and explains how the data is distributed among the clients. Section 3 details the baseline model architecture and the federated learning environment used in the experiments. The experimental evaluation and the analysis of the obtained results are presented in Sect. 4. Finally, Sect. 5 provides concluding remarks and outlines potential directions for future work.

2 NF-ToN-IoT Dataset

The NF-ToN-IoT dataset is a NetFlow-base dataset version of the ToN-IoT dataset, created by the University of New South Wales (UNSW) in Sydney, designed to support research in network intrusion detection in Internet of Things environments based on Artificial Intelligence [13]. It was developed at the UNSW Canberra Cyber IoT Lab and aims to emulate realistic and complex IoT environments, incorporating telemetry data from IoT sensors, operating systems logs, and network traffic data.

This dataset contains preprocessed and labeled network traffic data, captured from a simulated IoT network. The dataset includes both benign (19.6%) and malicious (80.4%) traffic, supporting a wide variety of attack types such as Denial of Service (DoS), Distributed DoS (DDoS), Ransomware, and Injection, among others (see Table 1). The data was generated using well-known cybersecurity tools like Wireshark, Zeek, Security Onion, and Kali Linux, adding reliability and realism to the scenarios represented.

Table 1. Distribution of traffic types in the NF-ToN-IoT dataset

Label	Number of Flows	Description
Benign	270,279	Normal unmalicious flows.
Backdoor	17,247	A technique that attacks remote-access computers by replying to specially constructed client applications.
DoS	17,717	Attempts to overload a system's resources to deny access or availability of its data.
DDoS	326,345	Similar to DoS, but originating from multiple distributed sources.
Injection	468,539	Attacks that supply untrusted inputs to alter program execution (e.g., SQL or code injections).
MITM	1,295	Man-in-the-Middle attack that intercepts communication between a victim and a host.
Password	156,299	Attacks aimed at retrieving passwords via brute force or sniffing techniques.
Ransomware	142	Encrypts a host's files and demands payment for the decryption key.
Scanning	21,467	Techniques that probe networks and hosts to discover information (also known as probing).
XSS	99,944	Cross-site scripting: attackers send malicious scripts via web applications to end users.

For this study, we focused on the network traffic data, which consisted of a CSV file representing all the captured packets along with their corresponding labels [6]. The dataset was preprocessed to extract the most relevant features, normalize the data, and transform the labels to indicate only whether a packet is malicious or not. First, the categorical Attack column was removed, as it was redundant for the binary classification task. Then, the feature matrix X was constructed by dropping the Label column, which was stored separately as the target vector y. Finally, all features were normalized using standard score normalization (z-score), ensuring that the input values had zero mean and unit variance.

The resulting dataset was divided into three main subsets: 70% for training, 15% for validation, and 15% for threshold selection (used to determine classification decision boundaries). A detailed breakdown of this partitioning is presented in Table 2 To simulate a federated learning environment, the training set was split evenly between two clients, allowing each to perform local training on distinct data partitions.

Table 2. Dataset partitioning and client distribution

Subset	Percentage	Number of Samples	Benign	Attacks
Training (Client 1)	35%	405,297	17.11%	82.89%
Training (Client 2)	35%	405,298	17.14%	82.86%
Validation	15%	173,699	17.28%	82.72%
Threshold Selection	15%	173,700	17.06%	82.94%
Total	100%	1,157,994		

3 Methodology

This section outlines the methodology followed to conduct the experiments and evaluate the proposed approach. First, we describe the baseline models used as a reference point to assess the effectiveness of our system. Then, we detail the federated learning environment, including the system architecture, data distribution strategy, and training procedures. Finally, we define the performance metrics used to measure and compare the models' behavior across various evaluation criteria.

3.1 Baseline

Before evaluating the effectiveness of federated learning in this context, it is necessary to define a baseline classification model. The chosen model is a Multi-Layer Perceptron (MLP) [5], a type of supervised learning algorithm that approximates a target function by training on labeled data. This model is well-suited for binary classification tasks, such as distinguishing between benign and malicious network traffic. It was implemented using TensorFlow/Keras [9] and trained on labeled NetFlow records.

The architecture of the model consists of an input layer with 10 neurons (one for each input feature), followed by two hidden layers with 64 and 32 neurons respectively, both using ReLU activation functions. The output layer consists of a single neuron with a sigmoid activation function to predict binary outcomes. The model was compiled using the Adam optimizer with a learning rate of 0.001, binary cross-entropy as the loss function, and accuracy as the evaluation metric. The input to the model consists of 10 features extracted from each NetFlow record. These features are summarized in Table 3.

3.2 Federated Learning Environment

To simulate a federated learning environment, we used the Flower framework [11], which allows seamless communication between multiple clients and a central server. Each client was assigned half of the training data, simulating local data ownership. Clients trained the same model architecture independently on their respective partitions and shared the resulting model weights with the

Table 3. Description of selected network flow features

Feature	Description
L4_SRC_PORT	Source port at Layer 4.
L4_DST_PORT	Destination port at Layer 4.
PROTOCOL	Network protocol used (e.g., TCP, UDP).
L7_PROTO	Application layer protocol.
IN_BYTES	Number of incoming bytes.
OUT_BYTES	Number of outgoing bytes.
IN_PKTS	Number of incoming packets.
OUT_PKTS	Number of outgoing packets.
TCP_FLAGS	Bitwise flags representing TCP control bits.
FLOW_DURATION_MILLISECONDS	Duration of the flow in milliseconds.

central server. The server aggregated these weights using the Federated Averaging (FedAvg) algorithm. This process was repeated for 12 global rounds, with each client performing 4 local training epochs per round. For comparison, the centralized model was trained for 20 epochs on the complete dataset.

The Federated Averaging (FedAvg) algorithm [14] combines model updates from multiple clients by computing a weighted average of their parameters. Each client performs local training on its data and then sends the updated model to the server, which aggregates the models based on the relative size of each client's dataset.

3.3 Performance Metrics

To assess and compare model performance, we employed standard classification metrics such as accuracy, loss, precision, recall, and F1-score [1,4], alongside visual tools like ROC curves [2] and confusion matrices [10]. These metrics provided insights into each model's ability to detect malicious traffic while minimizing false positives. Special attention was given to issues arising from class imbalance, where accuracy alone may be misleading. Binary cross-entropy loss was used to measure prediction error, while precision and recall helped evaluate the trade-off between false alarms and missed threats. The F1-score offered a balanced measure in imbalanced scenarios, and ROC curves, along with AUC values, summarized overall model effectiveness. The confusion matrix further enabled detailed analysis of prediction errors.

4 Results and Discussion

In this section, we present the results obtained from the evaluation of the federated learning approach for anomaly detection in IoT environments, using the NF-ToN-IoT dataset described in Sect. 2. For training, we used the base model defined in Sect. 3, configured for 12 federated rounds and 4 local epochs per client. Each client trained on data batches of size 32. The binary cross-entropy loss function was used, with a learning rate of 0.001. Accuracy was selected as the primary evaluation metric during training. The following Table 4 shows the main metrics compared between the baseline model (MLP) and the federated learning model (FedAvg).

Table 4. Comparison of model performance across centralized and distributed setups during evaluation

Setup	Accuracy	Loss	Precision	Recall	F1-Score	AUC
Centralized	0.9898	0.0273	0.9900	0.9977	0.9939	0.9992
Distributed	0.9913	0.0246	0.9922	0.9973	0.9948	0.9993

To gain a more comprehensive understanding of the model's performance, we evaluated it at each round using several classification metrics described in Sect. 3.3. In this analysis, we focus only on the server-side evaluation, as both the server and clients use the same evaluation dataset.

Initially, at round 0, the model's accuracy is very low due to its random weight initialization. However, after the first round of training, the model's accuracy increases significantly, stabilizing around 0.99 and reaching a final accuracy of 0.9913. Particular attention should be given to rounds 6 and 8, where a noticeable drop in accuracy occurs. This drop may be attributed to the way the FedAvg algorithm aggregates weights from different clients. In federated learning, the server combines the clients' local models through weighted averaging (FedAvg). When client models are poorly aligned or trained on non-IID data distributions [12], the aggregation can introduce instability, temporarily reducing model performance. Nonetheless, the model eventually recovers and continues to improve as training progresses.

In addition to accuracy, metrics like precision, recall, and F1-score provide a more detailed view of classification performance. As expected, at round 0, all metrics are low due to the model's lack of training. From round 1 onward, the model improves rapidly, with metrics surpassing 0.99 by rounds 5âĂŞ6, indicating robust classification capability. Similar to the accuracy trend, precision and recall drop to approximately 0.95 in rounds 6 and 8, reflecting the same instability caused by data heterogeneity. Following these dips, the metrics recover, suggesting that the model is adapting over time, despite the variability in data distributions (Fig. 1).

Once this stage is reached, we compare the performance of the federated and centralized models using key visualizations such as the evolution of training and validation loss and the confusion matrix. This analysis provides insights into model efficiency, convergence, generalization, and robustness, allowing us to assess the trade-offs of federated learning, particularly its advantages in privacy and decentralization, against traditional centralized approaches.

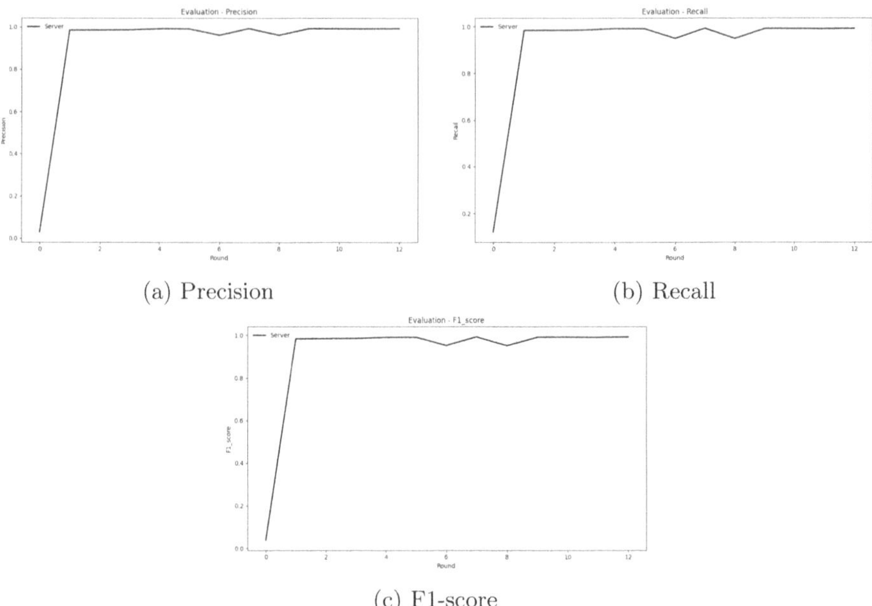

(a) Precision (b) Recall

(c) F1-score

Fig. 1. Precision, Recall and F1-score during federated evaluation across rounds.

The loss function is an essential indicator of model training. For each model, we plot both the training loss and validation loss across epochs for the centralized model, and across communication rounds for the distributed model. This allows us to visualize how the model's error evolves over time in each case.

In the centralized model graph 2, we can observe that the training loss and validation loss are very similar, with only a small difference between them. This is a good sign, as it suggests that the model is learning effectively during the training process and is able to generalize well to unseen data. The small gap between the two losses indicates that there is no significant overfitting, and the model is performing well throughout the epochs.

On the other hand, when we look at the distributed model, we notice a more interesting pattern. As we saw in the previous section, during rounds 6 and 8, there is a noticeable discrepancy between the training loss of the local models and the validation loss of the global model. This gap suggests that during these rounds, the local models were not perfectly aligned with the global model, possibly due to differences in the data each device was trained on. This misalignment can lead to a temporary drop in performance, as the global model struggles to incorporate the updates from the local models. However, it is important to note that this is a common issue in federated learning, and as more rounds progress, we expect the performance to stabilize and improve as the model continues to learn from the distributed data.

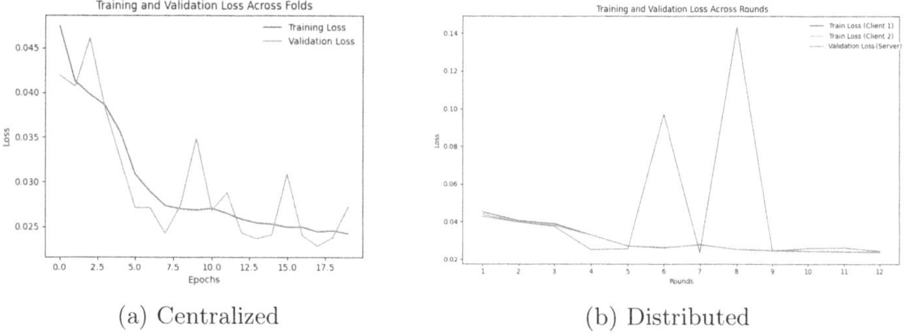

(a) Centralized (b) Distributed

Fig. 2. Training and Validation loss comparison.

Despite temporary fluctuations during training, the final outcomes are highly promising: both the centralized and federated models achieved an AUC-ROC of 0.9992, indicating exceptional discriminatory performance. To further assess classification efficacy, we examine the confusion matrices of both models, which provide a detailed breakdown of true negatives, false positives, false negatives, and true positives. This analysis helps validate that the federated approach retains performance comparable to the centralized model, even under decentralized training conditions.

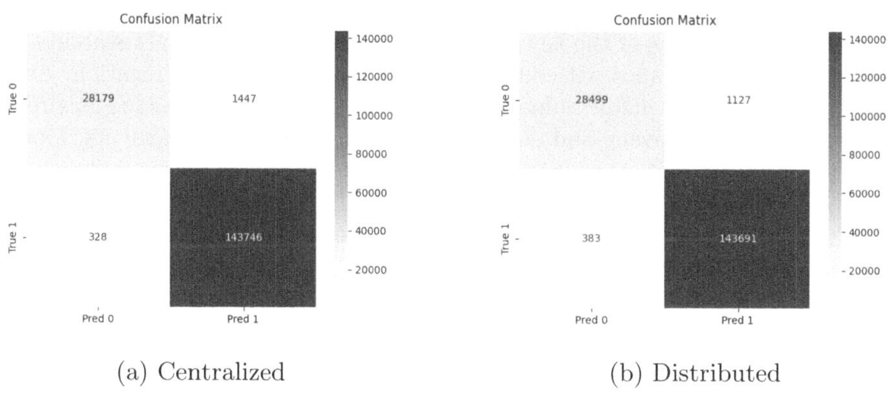

(a) Centralized (b) Distributed

Fig. 3. Confusion matrix comparison.

As shown in Fig. 3, the centralized model demonstrates excellent performance in accurately identifying attack traffic. However, when compared to the confusion matrix of the federated model, we observe that the centralized approach tends to classify more benign traffic as malicious, resulting in a higher number of false positives. From the confusion matrix, we obtain a precision of 0.9900, a recall of 0.9977, and an F1-score of 0.9939. These values suggest that the centralized model is highly effective at detecting attacks (high recall), while maintaining a relatively low false positive rate (high precision). The F1-score reflects a strong balance between these two aspects.

On the other hand, the federated model is able to slightly increase the precision to 0.9922, while still maintaining a recall of 0.9973. This results in an improved F1-score of 0.9948, indicating that the federated model slightly outperforms the centralized one in terms of overall classification performance. In particular, it reduces false alarms while continuing to detect nearly all attacks.

In conclusion, both models achieve outstanding performance, with the federated model matching or even surpassing the centralized model. While the performance gap between the two approaches is relatively small, these results highlight the potential of federated learning to provide similar (or even superior) outcomes to centralized training, all while preserving data privacy—a crucial advantage in security-sensitive domains such as IoT.

5 Conclusions and Future Work

To conclude this analysis, the federated model has demonstrated performance comparable to that of the centralized model, even outperforming it in certain evaluation metrics. Although the federated model required nearly twice as many training epochs as the centralized one, this extended training enabled the model to learn from multiple datasets distributed across different locations, all while preserving the privacy of each data source—an essential aspect when working with sensitive datasets in neural network training.

This highlights one of the key advantages of federated learning: its ability to achieve high performance without requiring centralized data aggregation. As a result, it represents a highly valuable solution for scenarios such as IoT environments, where data privacy and decentralization are crucial requirements. Moreover, its practical application is particularly useful in real-world contexts where different entities—such as hospitals, companies, or government institutions—prefer not to share sensitive data due to privacy concerns, yet aim to collaborate in training AI models for attack detection.

Future research on federated learning for intrusion detection in IoT should focus on improving robustness, scalability, and practical deployment. Analyzing the impact of data distribution (IID vs. non-IID) using synthetic datasets can help address performance issues from heterogeneity. Increasing the number of clients would enhance realism and test system scalability. Incorporating richer and more recent datasets could improve detection of sophisticated and evolving threats. Reducing communication overhead is also critical, especially in constrained environments. Techniques like model compression, gradient sparsification, and asynchronous updates offer promising solutions.

Acknowledgments. This publication is part of the AI4SECIoT project ("Artificial Intelligence for Securing IoT Devices"), funded by the National Cibersecurity Institute (INCIBE), derived from a collaboration agreement signed between the National Institute of Cybersecurity (INCIBE) and the University of Burgos. This initiative is carried out within the framework of the Recovery, Transformation and Resilience Plan funds, financed by the European Union (Next Generation), the project of the Government of Spain that outlines the roadmap for the modernization of the Spanish economy, the recovery of economic growth and job creation, for solid, inclusive and resilient economic reconstruction after the COVID19 crisis, and to respond to the challenges of the next decade.

Disclosure of Interests. The authors have no competing interests to declare that are relevant to the content of this article.

References

1. Classification: Accuracy, recall, precision, and related metrics. https://developers.google.com/machine-learning/crash-course/classification/accuracy-precision-recall. Accessed 07 May 2025
2. Classification: ROC and AUC. https://developers.google.com/machine-learning/crash-course/classification/roc-and-auc. Accessed 07 May 2025
3. Federated Learning. https://federated.withgoogle.com. Accessed 18 Jan 2025
4. Linear regression: Loss. https://developers.google.com/machine-learning/crash-course/linear-regression/loss. Accessed 07 May 2025
5. Neural network models (supervised). https://scikit-learn/stable/modules/neural_networks_supervised.html. Accessed 06 May 2025
6. NF-ToN-IoT-00-Cleaning. https://kaggle.com/code/dhoogla/nf-ton-iot-00-cleaning. Accessed 13 May 2025
7. Number of connected IoT devices growing 14% to 21.1 billion. https://iot-analytics.com/number-connected-iot-devices/. Accessed 04 May 2025
8. Riskiest Devices 2025 Report. https://www.forescout.com/resources/riskiest-devices-2025-report/
9. TensorFlow. https://www.tensorflow.org/. Accessed 06 May 2025
10. Thresholds and the confusion matrix. https://developers.google.com/machine-learning/crash-course/classification/thresholding. Accessed 07 May 2025
11. Authors, T.F.: Flower: a friendly federated AI framework. https://flower.ai/. Accessed 06 May 2025
12. Bhadauria, Y.S.: An introduction to Non-IID data and the challenges of data heterogeneity (2024). https://medium.com/@yuvraj.s.bhadauria/an-introduction-to-non-iid-data-and-the-challenges-of-data-heterogeneity-035baaaa5939. Accessed 07 May 2025
13. Luay, M., Layeghy, S., Hosseininoorbin, S., Sarhan, M., Moustafa, N., Portmann, M.: Temporal analysis of netflow datasets for network intrusion detection systems (2025). https://doi.org/10.48550/arXiv.2503.04404, http://arxiv.org/abs/2503.04404, arXiv:2503.04404 [cs]
14. McMahan, H.B., Moore, E., Ramage, D., Hampson, S., Arcas, B.A.Y.: Communication-efficient learning of deep networks from decentralized data (2023). https://doi.org/10.48550/arXiv.1602.05629, http://arxiv.org/abs/1602.05629, arXiv:1602.05629 [cs]
15. Sarhan, M., Layeghy, S., Moustafa, N., Portmann, M.: NetFlow datasets for machine learning-based network intrusion detection systems, vol. 371, pp. 117–135 (2021). https://doi.org/10.1007/978-3-030-72802-1_9, http://arxiv.org/abs/2011.09144, arXiv:2011.09144 [cs]

Author Index

E. Corchado et al. (Eds.): CISIS 2025, CCIS 2807, pp. 191–192, 2026.
https://doi.org/10.1007/978-3-032-19770-2

The manufacturer's authorised representative in the EU is Springer Nature Customer Service Centre GmbH, Europaplatz 3, 69115 Heidelberg, Germany. If you have any concerns regarding our products, please contact ProductSafety@springernature.com

Printed and bound by CPI Group (UK) Ltd, Croydon, CR0 4YY

07/07/2026

02160906-0002